OXFORD REVISION GUID

GCSE

BIOLOGY

through diagrams

W R Pickering

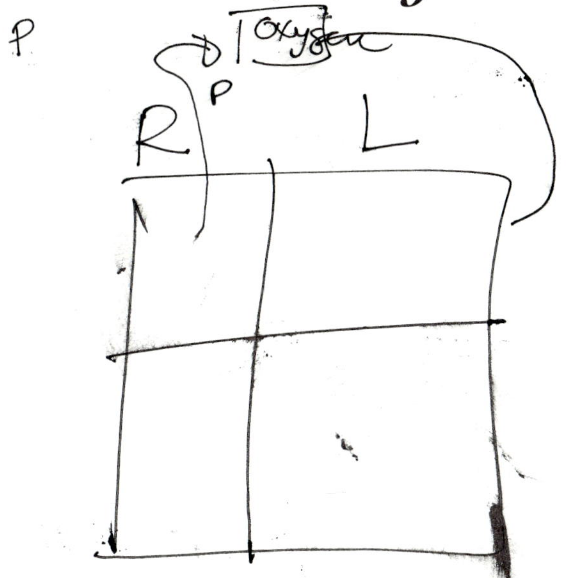

Oxford University Press

Oxford University Press,
Great Clarendon Street, Oxford OX2 6DP

Oxford New York
Athens Auckland Bangkok Bogota Bombay
Buenos Aires Calcutta Cape Town Dar es Salaam Delhi
Florence Hong Kong Istanbul Karachi
Kuala Lumpur Madras Madrid Melbourne
Mexico City Nairobi Paris Singapore
Taipei Tokyo Toronto Warsaw
and associated companies in
Berlin Ibadan

Oxford is a trade mark of Oxford University Press

First published 1996

New edition 1998 ISBN 0 19 914711 6 (School edition)
0 19 914712 4 (Bookshop edition)

Typesetting, design and illustration by Hardlines, Charlbury, Oxford
Printed in Great Britain

CONTENTS

Syllabuses for GCSE Science: Biology iv
Question types vi
Coursework vii
Success in examinations viii
How to revise ix
Revision timetable x

LIFE PROCESSES

Aerobic respiration 5
Human body systems 6
Angiosperm adaptations 7
Plant and animal cells 8
Diffusion, osmosis and active transport 9
Enzymes control biological processes 10

HUMAN PHYSIOLOGY

An ideal human diet 11
Malnutrition 12
Digestion, absorption and processing of foods 13
Mammalian circulation 14
Blood structure and function 15
Disease may have a number of causes 16
Food preservation 17
Natural defence systems of the body 18
Antibodies are specific protein molecules 19
Abuse of drugs 20
Neurones (nerve cells) carry information through the nervous system 21
The endocrine system 22
The eye is a sense organ 23
The brain is an integrator 24
Blood sugar regulation 25
Excretion 26
Kidney structure and function 27
Kidney failure: dialysis and transplants 28
Temperature control in endothermic organisms 29
Muscle-bone machines 30
Human reproduction 31
The menstrual cycle 32
The placenta 33

PLANT PHYSIOLOGY

Plants make food by photosynthesis 34
Transport systems in plants 35
Water movement through a plant 36
Hormones and minerals affect plant growth and development 37
Flower structure is adapted to pollination 38
Fertilisation, fruits and seed dispersal 39

VARIATION, INHERITANCE AND EVOLUTION

Keys and classification 40
Naming and classifying living organisms 41
Cell division and the human life cycle 42
DNA and chromosomes 43
DNA, genes and proteins 44
Cystic fibrosis and monohybrid inheritance 45
Sex linkage and the inheritance of sex 46
Variation, natural selection and evolution 47

BIOTECHNOLOGY AND GENETIC ENGINEERING

Genetic engineering (recombinant DNA technology) 48
Gene transfer can promote desirable characteristics 49
Cloning 50
Bioreactors/fermenters 51
Penicillin is an antibiotic 52
Bacteria and food production 53
Anaerobic respiration 54

ECOLOGY AND ECOSYSTEMS

Ecology 55
Factors affecting population growth 56
Human population growth 57
Pollution of the atmosphere 58
Pollution of water 59
Saprotrophs cause decay 60
Treatment of sewage 61
Ecological pyramids 62
Biological pest control 63
Cycling of nutrients 64
Managing ecosystems: fish farming 65
Managing ecosystems: horticulture 66
Managing ecosystems: animal husbandry 67
Design an experiment 68
Dealing with data 69
Graphical representation 70

Sample exam question answers 71
Self-assessment questions 75
Answers 79

INDEX 80

Syllabuses for GCSE Science: Biology

The subject of Biology may be studied in the following ways at GCSE:

- with Physics and Chemistry as part of **GCSE Science: Single Award** (equivalent to one GCSE)
- with Physics and Chemistry as part of **GCSE Science: Double Award** (equivalent to two GCSEs)
- on its own to give **GCSE Science: Biology** (one GCSE). If you are also studying Physics and Chemistry as separate subjects, this may lead to a triple award in the sciences.

The following information relates to syllabuses for GCSE Science: Biology.

Core and extension - two areas of Biology

GCSE Science: Biology covers some subject matter which is called the common core, and also extra areas of biology called extension material. The common core is what makes up the biological component of GCSE Science: Double Award.

Foundation and higher - two levels of examination

The scale of grades at GCSE is A*, A, B, C, D, E, F, and G. It would be impossible to grade pupils from A* to G based on the same examination, as pupils expecting an A would find all the questions too easy, or pupils expecting an F would find them too hard. To overcome this problem, there are two tiers or levels of exam called Foundation and Higher. Pupils entered for the Foundation tier can achieve grades G to C, and those entered for the Higher tier can achieve grades A* to D.

Examining groups and syllabuses - which one are you studying?

There are several examining groups or boards that set syllabuses and examinations, which are listed below with their abbreviations. Make sure you know which group and syllabus you are studying for. You can obtain a copy of the syllabus and also sample examination papers from your examining group, which will give you a good idea of what to expect when your examination comes. Full details of how to contact the examining groups are given on page viii.

EDEXCEL: London Examinations

MEG: Midland Examining Group

NEAB: Northern Examinations and Assessment Board

NICCEA: Northern Ireland Council for the Curriculum, Examinations and Assessment

SEG: Southern Examining Group

WJEC: Welsh Joint Education Committee

IGCSE: International General Certificate of Secondary Education, administered by University of Cambridge Local Examinations Syndicate

As well as the ordinary GCSE in Biology, MEG also offers two alternatives: Nuffield and Salters'. The Nuffield approach is based on learning through practical investigations, while the Salters' approach starts with everyday situations to introduce scientific ideas in a realistic context.

Assessment - how they award your grade

Your GCSE grade is based on two kinds of assessment: written **examinations** and **coursework**. There is one set of examinations for the Foundation tier and another set for the Higher tier, so you will sit either one set or the other. The examinations aim to test:

- your **knowledge** and **understanding** of biology - whether you can recall and use the subjects you have studied, and apply your knowledge to new situations
- whether you can **communicate** biological information to other people
- whether you can **evaluate**, or take in and make judgements on, information that is presented to you.

Experimental and investigative work is assessed by coursework for both the Foundation and Higher tiers. Coursework is explained in more detail on page vii.

In general, the examinations account for 75% of the final grade and the coursework for 25%. Some examining groups set two examinations, usually with the first one covering just the core material and the second also including extension material. Other examining groups set just one examination. The details of the assessment structure for each examining group and syllabus are given in the facing table. Find your own syllabus on the table and make sure you know what you are expecting. The 'Weighting' column shows what percentage of the total marks is given for each paper or for the coursework. There is more information about the sort of questions that may be set in examinations on page vi.

Assessment structure

Examination group/syllabus	Tiering	Papers	Length of paper	Weighting	Question style (questions compulsory?)
MEG					
Biology Syllabus A (1780)	Foundation G-C	Paper 1 (Core)	$1\frac{1}{2}$ hours	50%	Structured (all compulsory)
		Paper 3 (Extension)	$\frac{3}{4}$ hour	25%	Structured (all compulsory)
		Coursework		25%	
	Higher D-A*	Paper 2 (Core)	$1\frac{3}{4}$ hours	50%	Structured (all compulsory)
		Paper 4 (Extension)	1 hour	25%	Structured (all compulsory)
		Coursework		25%	
Biology Syllabus D (Nuffield) (1785)	Foundation G-C	Paper 1 (Core)*	$1\frac{1}{2}$ hours	50%	Structured (all compulsory)
		Paper 3 (Extension)	$\frac{3}{4}$ hour	25%	Structured (all compulsory)
		Coursework		25%	
*same as Syllabus A	Higher D-A*	Paper 2 (Core)*	$1\frac{3}{4}$ hours	50%	Structured (all compulsory)
		Paper 4 (Extension)	1 hour	25%	Structured (all compulsory)
		Coursework		25%	
Biology Syllabus C (Salters') (1790)	Foundation G-C	Foundation Paper	2 hours	75%	Structured (all compulsory)
		Coursework		25%	
	Higher D-A*	Higher Paper	$2\frac{1}{4}$ hours	75%	Structured (all compulsory)
		Coursework		25%	
NEAB					
Biology (1161)	Foundation G-C	Paper F	2 hours	75%	Structured (all compulsory)
		Coursework		25%	
	Higher D-A*	Paper H	$2\frac{1}{4}$ hours	75%	Structured (all compulsory)
		Coursework		25%	
EDEXCEL					
Biology (1026)	Foundation G-C	Paper 1F (01) (Core)	$1\frac{1}{2}$ hours	50%	Structured (all compulsory)
		Paper 2F (02) (Extension)	1 hour	25%	Structured (all compulsory)
		Coursework		25%	
	Higher D-A*	Paper 3H (03) (Core)	$1\frac{1}{2}$ hours	50%	Structured (all compulsory)
		Paper 4H (04) (Extension)	1 hour	25%	Structured (all compulsory)
		Coursework		25%	
SEG					
Biology (2604)	Foundation G-C	Core paper (2)	$1\frac{1}{2}$ hours	50%	Structured and free response (all compulsory)
		Additional paper (4)	1 hour	25%	Structured and free response (all compulsory)
		Coursework		25%	
	Higher D-A*	Core paper (3)	$1\frac{1}{2}$ hours	50%	Structured and free response (all compulsory)
		Additional paper (5)	1 hour	25%	Structured and free response (all compulsory)
		Coursework		25%	
NICCEA					
Biology	Foundation G-C	Paper 1	1 hour	30%	Short answer
		Paper 2	$1\frac{1}{2}$ hours	45%	Structured (all compulsory)
		Coursework		25%	
	Higher (E)D-A*	Paper 1	$1\frac{1}{2}$ hours	32%	Short answer
		Paper 2	2 hours	43%	Structured (all compulsory)
		Coursework		25%	
WJEC					
Biology	Foundation G-C	Foundation paper	2 hours	75%	Short answer and free response
		Coursework		25%	
	Higher D-A*	Higher paper	$2\frac{1}{2}$ hours	75%	Short answer and free response
		Coursework		25%	
IGCSE					
Biology	Core G-C	Paper 1	$\frac{3}{4}$ hour	40%	Multiple choice (compulsory)
		Paper 2	1 hour	40%	Short answer and structured
		Coursework			
		Paper 4 coursework *or*		20%	
		Paper 5 Practical test *or*	1 hour	20%	Questions covering experimental and observation skills
		Paper 6 Alternative to practical	1 hour	20%	Written questions about laboratory procedures
	Extended A to G	Core Papers 1 and 2 *plus* Paper 3	$1\frac{1}{4}$ hours	80%	Structured and free response
		Coursework			
		Paper 4 coursework *or*		20%	
		Paper 5 Practical test *or*	1 hour	20%	Questions covering experimental and observation skills
		Paper 6 Alternative to practical	1 hour	20%	Written questions about laboratory procedures

Question types

Your examination may contain several of the same type of question, or it may contain different styles of question. The table on the previous page shows the question style set by each examining group. The main types of question you may come cross are described below. Some sample examination style questions are given on page 71.

Short-answer questions

These questions can usually be answered in one sentence, or with a drawing or equation. The answer is written in a space left on the examination paper, and the number of marks is given.

Structured questions

These are longer questions made up of connected parts. They are presented in the same way as short-answer questions, the main difference being that they enlarge on a topic in more detail. The number of marks is given on the paper for each part of the question.

Answering short-answer and structured questions

The space left on the paper gives you an idea of how much to write, and you should keep your answer brief and concise. Make sure you read the question carefully and don't give information that is not required. For example, if you are asked to list then do just that - don't describe or explain or discuss, you will waste time and you won't gain any more marks. The number of marks also gives a good indication of how much the examiner is looking for. If there are three marks for a 'list' question, then there will probably be three things to list.

If you are asked to sketch a diagram, then don't waste time making it look beautiful. It should be neat, clear and accurately labelled but not elaborate.

Free response questions

These are open-ended questions which require you to build up an explanation or argument in an essay-type answer. Free response questions are only set by a few of the examining groups. One question like this can carry quite a lot of marks.

Answering free response questions

You need to judge how much information to include, and how much you should write. It's a good idea to plan your answer to this type of question first, by making notes or sketches to organise your thoughts.

Make sure you read the question very carefully, because if you go off in the wrong direction you will lose a lot of marks. Use the number of marks to decide how much time you should spend on the question. If the question is worth 8 marks out of a total of 24, then don't spend more than about a third of the total time on it.

Try to be as neat and as logical as you can. This will help the examiner to find his or her way around your answers, and to award the marks quickly. Don't forget that spelling, punctuation, and grammar are important, and you will lose marks if your written English is sloppy.

Multiple choice questions (IGCSE only)

In these questions, several alternative answers are suggested and you have to choose which one is correct. The examination paper requires you to fill in your answer in a specific way, and you will lose marks if you don't record your answer correctly. You may lose marks for an incorrect answer, so if you are really not sure it might be best to leave that question blank. The best preparation for this kind of question is to look at sample/past papers from your examining group and read the instructions carefully.

Don't write too much! Just answer the questions asked.

Coursework

Practical work, including investigations, will be marked by your teacher throughout the course and will count 25% towards your final grade (If you are taking the IGCSE exam then the arrangements are different - check with your teacher.)

There are four separate skill areas in which you will be awarded a mark:

- planning experimental procedures (P)
- obtaining evidence (O)
- analysing the evidence and drawing conclusions (A)
- evaluating the evidence (E).

These four areas are explained in more detail below. In most syllabuses you will be awarded a mark out of 8 or out of 6 in each area, with a small additional mark for spelling, punctuation, and grammar. Your teacher will tell you when you are going to be tested, and on which area or areas. You will have opportunities throughout the course to practise and improve your skills. Your best mark in each of the four areas will normally be used, though the marks must come from at least two different pieces of coursework, and one of them should come from a complete investigation. This means that instead of just carrying out an experiment according to instructions given to you, you are given a problem to solve. You have to decide what you are going to look at, what measurements you will take, how to measure them, how to present your results, and finally assess whether you have actually solved the problem or not.

Planning experimental procedures (P)

First read the problem carefully and think about what you are trying to find out. Then write a short plan including:

- the problem you are trying to solve
- the variables you will keep constant and the variable you will change, and any control experiments. For example, if you are testing whether Benedict's solution changes colour when heated with a solution of glucose, you need to make sure the Benedict's solution does not change colour when heated without glucose. Your control experiment might be to heat the Benedict's solution with the same volume of water, instead of with the glucose solution.
- the apparatus you will need, perhaps with a labelled diagram of the set-up, and any safety measures you will need to take
- what you will measure, and how many readings you need to take
- your predicted results, such as: 'If I change this variable, then I predict that that will happen, because...'.

Obtaining evidence (O)

Here you are tested on your ability to carry out practical work. Follow your plan, or other instructions given to you, step by step and use apparatus carefully to make accurate measurements. Observe what is happening very carefully. If your planned experiment does not work for some reason, you can make changes to your method.

Record all your results carefully and clearly, either in a list or in a table with precise headings. Remember to include units. Repeat the experiment if you can, to obtain a second set of results. Repeat any readings you think might be inaccurate.

Analysing the evidence and drawing conclusions (A)

You need to present your list or table of results in a form that shows clearly what the experiment demonstrated. A graph is often (but not always) the best way of doing this. Put the variable that you changed along the x-axis, and the variable that you measured up the y-axis. For example, if you measured the number of bubbles of oxygen given off by pondweed at different distances from a light source, then the distance from the light source (which you determined by placing the tubes at different positions) goes along the x-axis, and the number of bubbles (which you measured) goes up the y-axis. Label the axes and don't forget units. If you have results that don't seem to fit, you should still include them. You may be able to think of a reason why they are inaccurate.

Then draw conclusions. This means describing what your results actually show, not what you expected them to show! Comment on the shape of your graph and what this means. Explain your conclusions using biological knowledge.

Evaluating the evidence (E)

You need to decide how well the investigation went. This account should include answers to the following questions:

- Were your results accurate? Do they agree with other results for the same experiment?
- Were your results reliable? If you repeated the experiment, did you get similar results each time?
- Do your results agree with your prediction? If not, is this because the prediction was wrong, or because of a fault in the design of the method?
- How would you improve your experiment if you were to do it again?

Success in examinations

If you have worked hard for two years and put in a lot of effort, there are some steps you can take to ensure that your examination result reflects this.

Be prepared

If the thought of an examination fills you with dread, then you are probably not fully prepared. Of course everybody will be a bit nervous about an important examination, but there is no reason why you should fall apart and fail to do yourself justice if you know what to expect before you go in. The best preparation is working through past papers and looking at your syllabus to avoid any nasty surprises. A positive attitude will help you - try to look on the examination as an opportunity to show the examiner all the things you have learnt in the past two years.

Make sure you have everything you need ready the night before, including pens, pencils, ruler, and calculator. Your teacher can tell you what mathematical skills you are expected to have, or they are listed in your syllabus. Make sure you are familiar with these well in advance.

Read carefully

Before you start, read the paper all the way through and make sure you know exactly what you have to do. Turn over all the pages. Then, when you are ready to start, read each question more than once to make sure it really does say what you think it says. Follow the instructions to the letter, as marks can only be awarded for precise answers to the question asked, not for any other information about the subject.

Plan your time

Work out how much time you should spend on each question, based on how many marks it has. Don't spend ages racking your brains to think of the answer to the last part of a question - go on and start the next question instead. You will probably get more marks for answering the beginning of all the questions than for finishing one or two questions completely, but not attempting later questions. If you have time left at the end, you can go back and think about the more difficult parts when you will be feeling more relaxed.

Present your work clearly

Remember that the examiner will have a large pile of papers to mark in a short space of time. Struggling to read your writing or to follow an illogical argument will not help, so write as clearly as you possibly can in the time available and think through what you are going to write before you start. The examiner is trying to award you marks - make it easy to find them.

If you are drawing diagrams, draw them clearly and simply, using single lines where appropriate. Label the diagrams and make sure the label lines point exactly to the relevant structures.

Stay calm

If you find a question you have no idea about, do not panic! Breathe slowly and deeply and have another look. It's likely that if you stay calm and think clearly, it will start to make more sense, or at least you may be able to answer part of it. If not, then don't agonise about it but concentrate first on the questions you can answer.

Exam Organisation Addresses

EDEXCEL (formerly ULEAC)
Stewart House, 32 Russell Square, LONDON WC1B 5DN
Tel: 0171 331 4000 Fax: 0171 331 4044

Midland Examining Group (MEG) and International General Certificate of Education (IGCSE)
Head office, Syndicate Buildings, 1 Hills Road, CAMBRIDGE CB1 2EU
Tel: 01223 553311 Fax: 01223 460278

Northern Examining and Assessment Board (NEAB)
Head office, Devas Street, Manchester M15 6EX
Tel: 0161 953 1180 Fax: 0161 273 7572

Northern Ireland Council for the Curriculum, Examinations and Assessment (CCEA)
Beechill House, 42 Beechill Road,
BELFAST BT8 4RS
Tel: 01232 704666 Fax: 01232 799913

Southern Examining Group (SEG)
Associated Examining Board, Stag Hill House,
GUILDFORD, Surrey GU2 5XJ
Tel: 01483 506506 Fax: 01483 300152

Welsh Joint Education Committee (WJEC)
Welsh Joint Education Committee, 245 Western Avenue,
CARDIFF CF5 2YX
Tel: 01222 265000 Fax: 01222 575895

How to revise

You will probably have found by talking to other pupils that everybody has a different way of revising. It's important to make sure that your method is an effective one.

Make a revision timetable

First plan your revision. You have a lot of subjects to fit in, but you can cover them all, and keep up with your hobbies too, if you plan your time carefully. A blank revision timetable is provided on page x. Make a timetable that includes all your subjects, and think about combining the subjects you find difficult with those that come more naturally to you, so you have something to look forward to and don't get disheartened. Use your syllabus to make sure you have covered everything. Once you've made the timetable, though, make sure you really do start revising - don't kid yourself that planning your revision is all you need to do!

Find a suitable place to revise

You need somewhere quiet where you will not be interrupted. Some people can concentrate with music playing, but if you can't, then turn it off. If you find it difficult to work at home, then try going to a library or some other quiet place.

Work with the material

Don't just sit there reading. You need to interact or work with the main ideas to get them fixed in your mind, rather than just trying to learn reams of words on a page. Different people do this in various ways, but you could try the following:

- if this book is your own, using a highlighter pen to show key words
- making brief notes from a textbook, or reorganising your own notes
- organising the information in different ways, such as making lists or flow diagrams, or using coloured highlighters to code related information
- drawing pictures or making up mnemonics or rhymes to help you remember key facts or important ideas
- reading a section and then testing yourself.

Try working with a friend, and don't forget to ask your teacher for advice if you get stuck.

Take regular breaks

Your concentration will start to wander and you will be less effective if you work solidly for hours without a break. Divide up your study periods into chunks of about 40 minutes, and then go and make a drink, or have a brief walk, or do something completely different for about 10 minutes. Then you can go back for the next session feeling refreshed, and sometimes you will find that you understand things more clearly after a break.

Go over topics repeatedly

You can fix things in your memory most effectively if you go back over them regularly. Each time you start a revision session, briefly go over or test yourself on some of the things you have learnt earlier. The more times you think about a topic, the more likely you are to remember it, even when under pressure in an examination.

Work through past examination papers

It's important not only to learn the information, but also to be able to answer the sorts of questions that are likely to come up. There is more about this on page 71.

Don't panic at the last minute

If you have stuck to your revision timetable, you should feel as relaxed and confident as you can about the examination. There is then no point in panicking and trying to learn it all again the night before. Just have a quick look through your brief notes, diagrams or lists, test yourself on a couple of topics, and then relax. If you are nervous and feel convinced you have forgotten everything, then take some exercise or do something else to clear your thoughts and stop worrying. You will probably find that when you come back and have another look, you have remembered far more than you thought.

Revision Timetable

Date							
Day	**Monday**	**Tuesday**	**Wednesday**	**Thursday**	**Friday**	**Saturday**	**Sunday**
Subject/Topic							
Time/Comments							
Date							
Day	**Monday**	**Tuesday**	**Wednesday**	**Thursday**	**Friday**	**Saturday**	**Sunday**
Subject/Topic							
Time/Comments							
Date							
Day	**Monday**	**Tuesday**	**Wednesday**	**Thursday**	**Friday**	**Saturday**	**Sunday**
Subject/Topic							
Time/Comments							
Date							
Day	**Monday**	**Tuesday**	**Wednesday**	**Thursday**	**Friday**	**Saturday**	**Sunday**
Subject/Topic							
Time/Comments							
Date							
Day	**Monday**	**Tuesday**	**Wednesday**	**Thursday**	**Friday**	**Saturday**	**Sunday**
Subject/Topic							
Time/Comments							

Aerobic respiration involves the oxidation of food to release energy to perform work in living organisms.

EQUATION FOR AEROBIC RESPIRATION
and suggested methods for demonstrating this process

GLUCOSE + OXYGEN $\xrightarrow{\text{ENZYMES}}$ ENERGY + CARBON DIOXIDE + WATER

Oxygen consumption can be measured using a **respirometer**

Involvement of enzymes can be demonstrated by the effects of temperature on respiration.

Energy release can be measured as the production of **heat**.

Carbon dioxide release causes
- limewater to turn milky
- bicarbonate indicator solution to turn from red → yellow-orange.

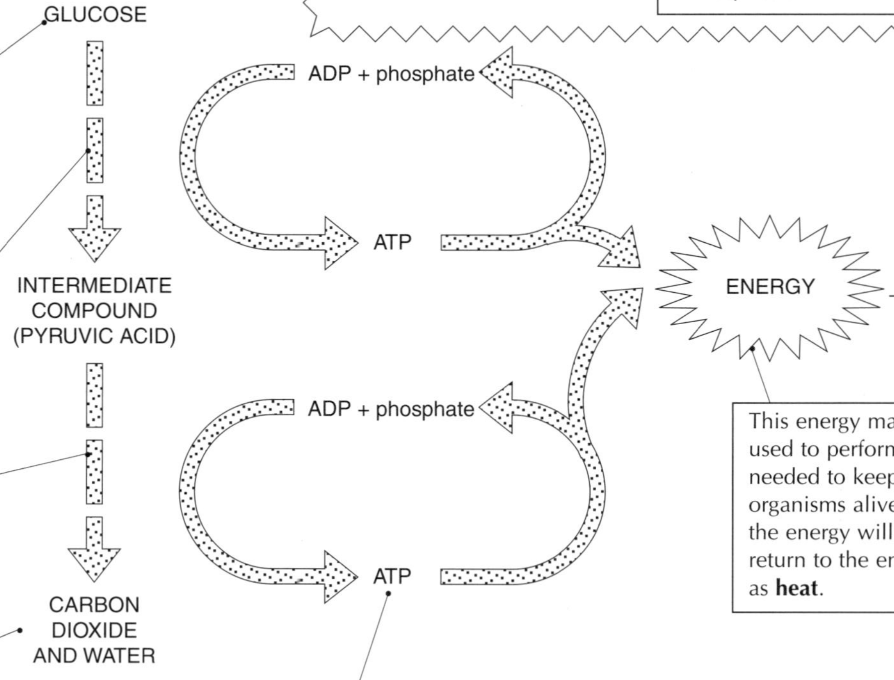

Obtained from food by the digestion of sucrose or starch, or from storage by breakdown of glycogen.

These reactions can take place **without oxygen** but release **very little ATP** (see below).

These reactions occur in the mitochondria of cells. They take place **only if oxygen is present** but release **large amounts of ATP**.

These end-products may be excreted (carbon dioxide through the lungs, for example). This **respiratory water** may be vital for desert animals which obtain very little liquid water in their diet.

This energy may now be used to perform the **work** needed to keep cells and organisms alive. Much of the energy will eventually return to the environment as **heat**.

ATP (Adenosine triphosphate) is a convenient form of 'energy currency'. Its availability means that
- the energy released by respiration is trapped in convenient packets
- the energy packets can be spent (when ATP breaks down to ADP - adenosine diphosphate - and phosphate) when and where it is needed.

Growth: involves the consumption of energy for the synthesis of large molecules from smaller ones (e.g. proteins from amino acids).

Maintenance of body temperature: all chemical reactions in the body release some heat. Reactions in liver and muscles are particularly important.

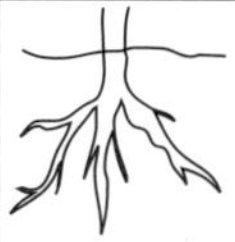

Active transport: the movement of molecules against a concentration gradient, e.g. ion uptake by roots.

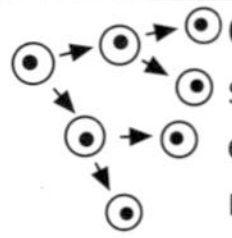

Cell division: replication and separation of chromosomes is very energy demanding. A human replaces billions of cells every day.

Movement: contraction of muscle requires energy. The supply of energy will run out if food or oxygen is limited.

Human body systems

all play a part in homeostasis.

HUMANS ARE MAMMALS.

Humans
- breathe through lungs
- have a diaphragm separating chest from abdomen
- fertilise eggs and develop their young inside the body of the female
- bear live young
- feed their young on milk from specialised mammary glands
- have some body hair, which may once have helped to make humans endothermic

** **Humans are extremely successful because they are homeostatic (can keep a 'steady state') and thus independent of their environment**

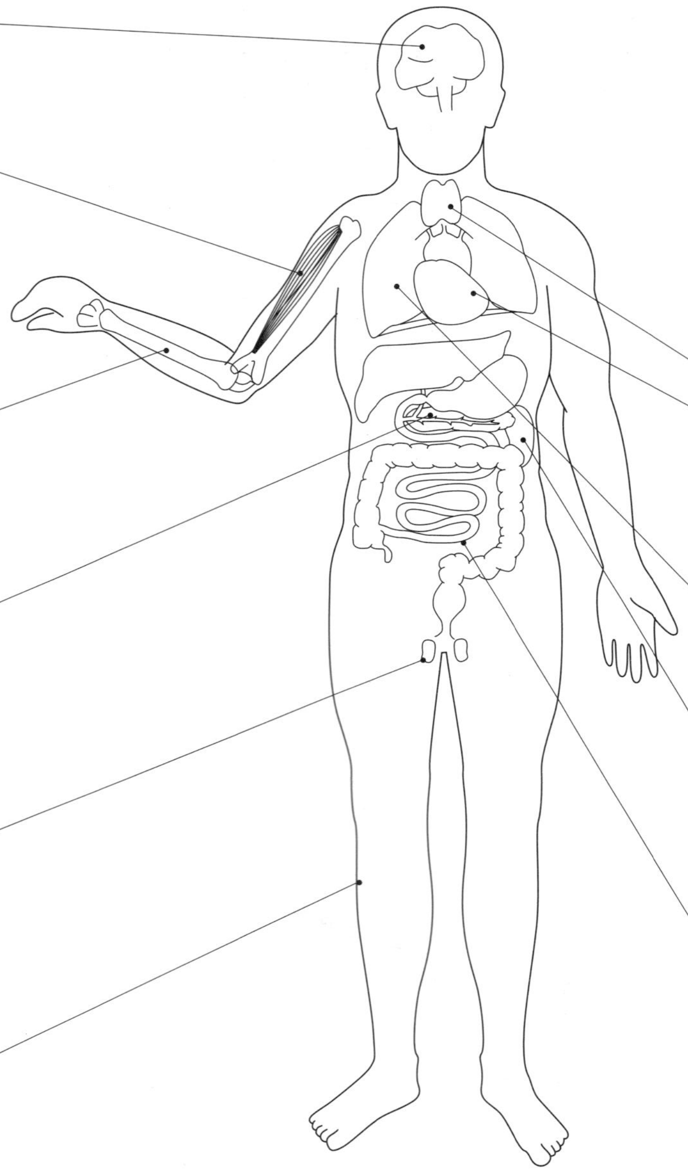

The nervous system is made up of **brain**, **spinal cord**, **sense organs** and **peripheral nerves**. Conducts impulses responsible for integration of other systems. The principal regulatory system.

The muscular system comprises **skeletal**, **cardiac** and **smooth muscle**. Responsible for the locomotion, movement of body parts, pumping of blood and internal movement of other material. Maintains blood pressure, responsible for formation of tissue fluid.

The skeletal system consists of **bones** and **cartilage**. Supports the body, protects soft tissues, site of calcium storage and blood cell synthesis.

The endocrine system consists of the **ductless glands**, many of which are under the influence of the pituitary gland. These regulate many body functions, and help to keep a constant composition of the blood.

The reproductive systems comprise **testes** (male), **ovaries** (female) and associated structures. Gamete production and transfer, and maintenance of secondary sexual characteristics.

Skin, **hair**, **nails** and **sweat glands** protect body against infection and dehydration, help control body temperature and receive stimuli such as pressure.

The circulatory system is composed of two subsystems. The **cardiovascular system** (heart, blood vessels and blood) is the transportation system responsible for distribution of many solutes, and the **lymphatic system** (lymph vessels and lymph) is responsible for return of tissue fluid to blood and for defence against diseases.

The respiratory system consists of the **lungs** and **air passageways** and maintains optimum concentrations of carbon dioxide and oxygen in the tissues.

The urinary system comprises the **kidneys**, **bladder** and associated **ducts**. These produce urine and remove it from the body. They remove toxins from, and maintain optimum solute concentrations in the blood.

The digestive system comprises **mouth**, **oesophagus**, **stomach**, **intestines** and accessory organs (principally **liver** and **pancreas**). These organs ingest food, break it down mechanically and chemically, and absorb nutrients so maintaining optimum concentrations of fuel molecules and raw materials for syntheses.

Angiosperm adaptations

- more than 80% of all plant species are angiosperms (i.e. plants with enclosed seeds).

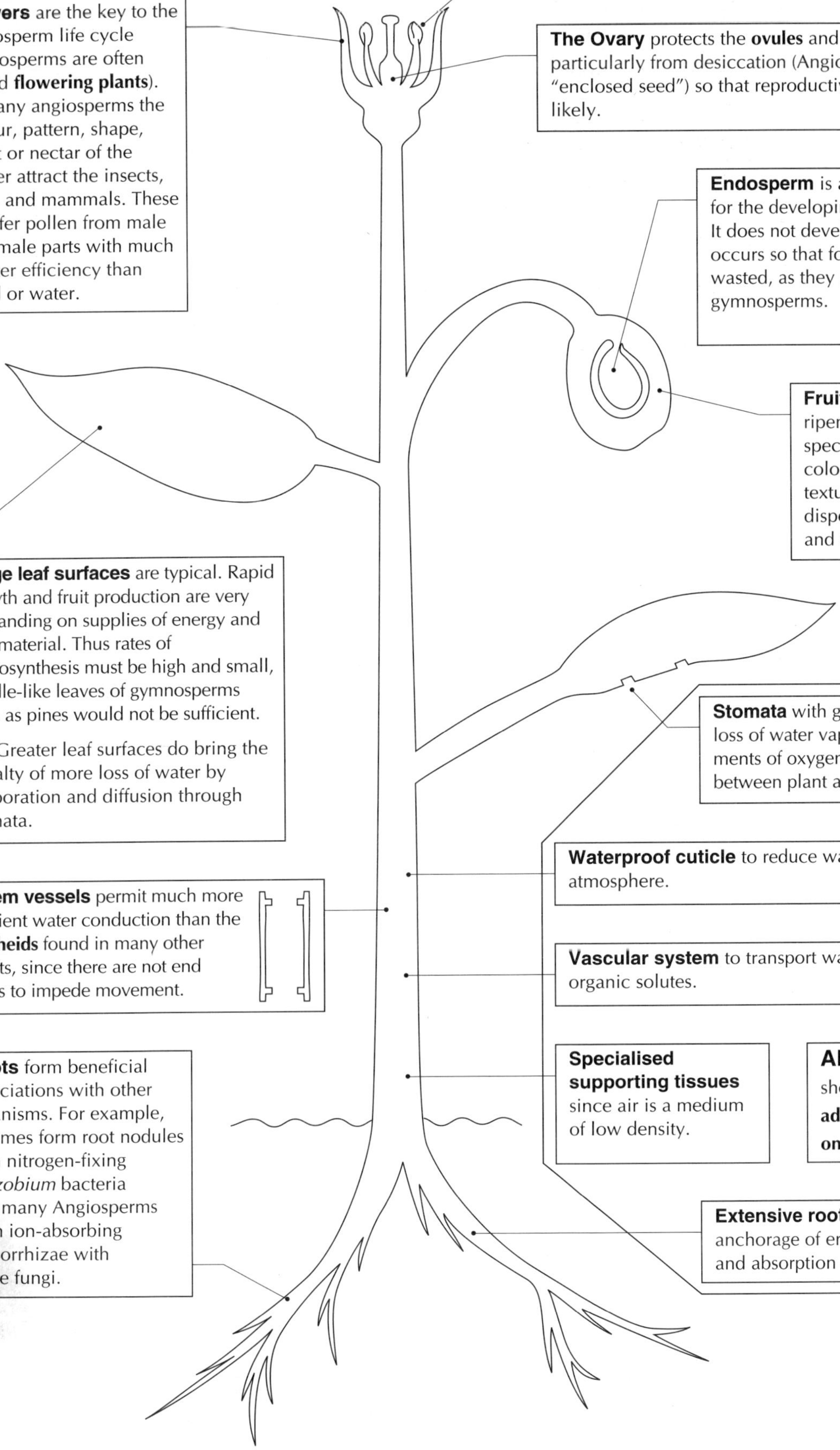

Growth, particularly of reproductive structures, can be very rapid. For example, pollination, pollen tube formation and fertilisation may take only one hour (compared to one *year* in some gymnosperms), and some bamboos can grow 1 metre per day.

Flowers are the key to the angiosperm life cycle (angiosperms are often called **flowering plants**). In many angiosperms the colour, pattern, shape, scent or nectar of the flower attract the insects, birds and mammals. These transfer pollen from male to female parts with much greater efficiency than wind or water.

The Ovary protects the **ovules** and **developing embryo**, particularly from desiccation (Angiosperm means "enclosed seed") so that reproductive success is more likely.

Endosperm is a store of nutrients for the developing plant embryo. It does not develop until fertilisation occurs so that food stores are not wasted, as they might be in gymnosperms.

Fruits are formed from ripened ovaries. Their specialised shapes, colours, smells and textures aid seed dispersal by wind, water and animals.

Large leaf surfaces are typical. Rapid growth and fruit production are very demanding on supplies of energy and raw material. Thus rates of photosynthesis must be high and small, needle-like leaves of gymnosperms such as pines would not be sufficient.

NB Greater leaf surfaces do bring the penalty of more loss of water by evaporation and diffusion through stomata.

Stomata with guard cells to regulate loss of water vapour and movements of oxygen and carbon dioxide between plant and atmosphere.

Waterproof cuticle to reduce water loss to atmosphere.

Xylem vessels permit much more efficient water conduction than the **tracheids** found in many other plants, since there are not end walls to impede movement.

Vascular system to transport water, ions and organic solutes.

Specialised supporting tissues since air is a medium of low density.

ANGIOSPERMS show basic plant **adaptations to life on land**

Roots form beneficial associations with other organisms. For example, legumes form root nodules with nitrogen-fixing *Rhizobium* bacteria and many Angiosperms form ion-absorbing mycorrhizae with some fungi.

Extensive root systems for anchorage of erect shoot systems and absorption of water and ions.

Plant and animal cells

ANIMAL CELL FEATURES

often relate to heterotrophic nutrition and high rates of metabolic activity.

Secretory vesicles containing cell products such as hormones or enzymes are much more common in animal cells.

Cytoplasm of animal cells is often denser, with many more organelles and dissolved substances.

Vacuoles are small and temporary. They can be involved with digestion (e.g. in phagocytes) or with excretion (contractile vacuoles may remove excess water).

Glycogen is the storage form of carbohydrates.

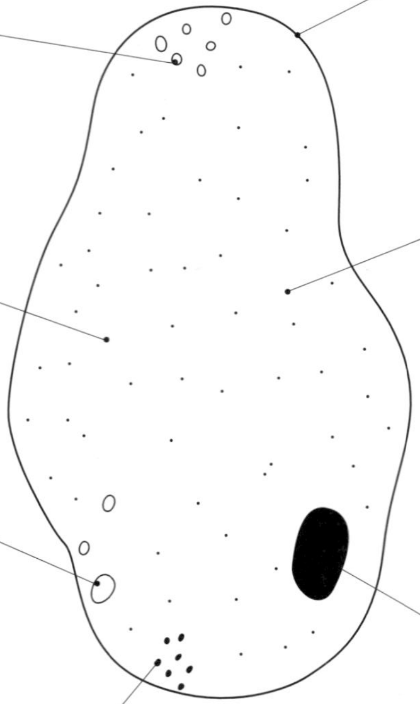

PLANT AND ANIMAL CELLS HAVE COMMON FEATURES which relate to maintaining the characteristics of life.

Cell membrane which surrounds the cytoplasm. It controls the **entry and exit** of dissolved substances and is therefore responsible for separating the cells contents from its surroundings.

Cytoplasm contains water, dissolved substances such as amino acids and sugars, and supports the various organelles (for example, mitochondria, ribosomes). It is within the cytoplasm and organelles that the various metabolic reactions needed to sustain life take place (for example, respiration).

Nucleus contains the genetic material (**DNA** which makes up **genes** or the **chromosomes**) which carries the coded instructions controlling the activities and characteristics of the cell. The chromosomes only become visible during cell division.

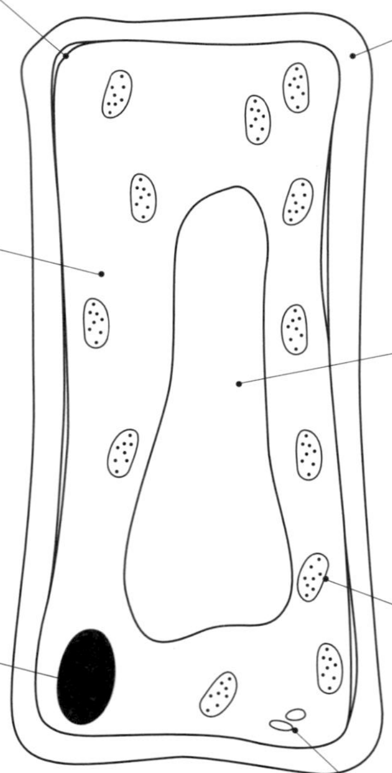

PLANT CELL FEATURES

often relate to autotrophic nutrition.

Cellulose cell wall provides structural support (pressure of cell contents leads to **turgidity**) and protects against damage caused by osmotic intake of water. It is **freely permeable to water and dissolved substances.**

Large permanent vacuole contains water necessary to provide turgor pressure and may be store for ions and molecules.

Chloroplasts contain the pigment **chlorophyll** (light absorption) and the **enzymes** necessary for the production of glucose by photosynthesis.

Starch (in the cytoplasm or the chloroplasts) is the storage form of carbohydrate.

The absence of the cellulose cell wall means that animals cells may be **very irregular in shape** and the limit to the amount of cytoplasm which can be controlled by the nucleus means that animal cells may be **quite small** - about 25 μ diameter.

The presence of the cellulose cell wall means that plant cells tend to be **regular in shape** and the presence of the vacuole means that plant cells may be **quite large** - often 60 μm (or 0.06 mm) in diameter.

Diffusion, osmosis and active transport are processes by which molecules are moved. Diffusion and osmosis are passive, but active transport requires energy.

DIFFUSION:
the movement of ions or molecules down a concentration gradient i.e. from a region of higher concentration to one of lower concentration.

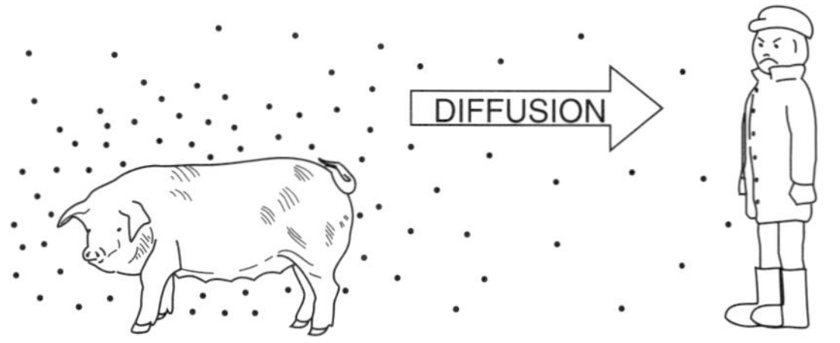

This is a physical process which depends on the energy possessed by the molecules, thus
- small molecules diffuse faster than large molecules
- diffusion speeds up as temperature increases.

For living cells the principle (the movement of molecules down a concentration gradient) is the same, but there is one problem → the cell is surrounded by a **cell membrane** which can restrict the free movement of the molecules

→This is **a selectively permeable membrane:** the composition of the membrane (lipid and protein) allows some molecules to cross with ease, but others with difficulty or not at all. In this example the membrane is permeable to water ● but not to the larger sucrose molecule (hatched circle) - the simplest sort of selection is based on the **size** of the molecules.

Important examples
- Oxygen from air sacs in the lung to blood, and from blood to cells
- Soluble foods from gut to blood
- Carbon dioxide from air to spaces inside leaf

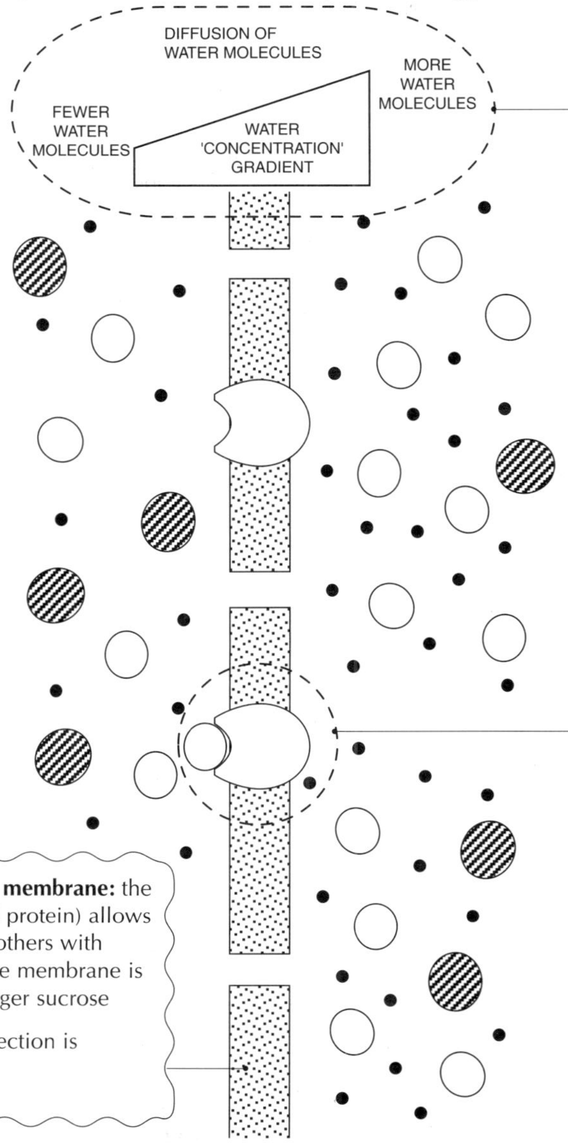

OSMOSIS IS THE DIFFUSION OF WATER

Water crosses membranes very freely and always tends to move, by diffusion, down the water 'concentration' gradient. The term 'concentration' can be confusing when used to describe water molecules, and is better replaced by the term 'potential'.

Thus osmosis is
- the movement of water
- across a selectively permeable membrane
- down a water potential gradient

Osmosis is responsible for water movement
- from tissue fluid to cells
- from soil water to root hairs
- from xylem to leaf mesophyll cells

ACTIVE TRANSPORT MAY MOVE MOLECULES AGAINST A CONCENTRATION GRADIENT

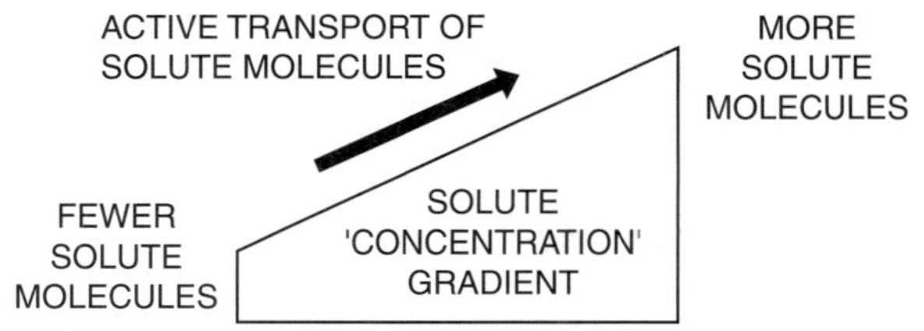

In this example there are more amino acid molecules ○ on the right side of the membrane than on the left - to move any more from left to right will be 'uphill', **against** the amino acid gradient. This active transport
- requires energy to 'drive' the molecules 'uphill' - this energy is supplied as ATP from respiration
- is affected by any factor which affects respiration, e.g. temperature and oxygen concentration
- is carried out by 'carrier proteins' in the membrane, which bind to the solute molecule, change shape, and carry the molecule across the membrane.

Important examples are
- uptake of mineral ions from soil by root hair cells
- movement of sodium ions to set up nerve impulses

Enzymes control biological process

and are widely exploited by humans.

TEMPERATURE: like all proteins, enzymes are made up of long, precisely folded chains of amino acids. This folding may be 'undone' by high temperature so that the enzyme may lose its active site - it is **denatured**.

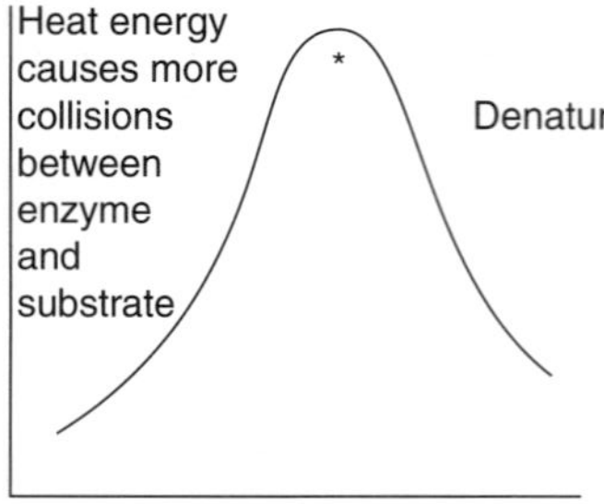

* The optimum temperature for human enzymes is close to 37°C. For most plants it is lower.

pH: Is a measure of acidity or alkalinity, and is a mathematical method for expressing the concentration of H^+ ions in a solution.

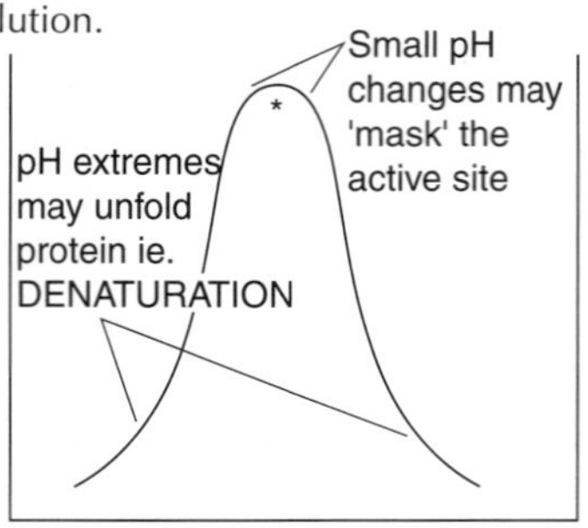

* the optimum pH for an enzyme depends on its site of action e.g. enzymes in the stomach (where HCl is present) have an optimum about pH2 but intestinal enzymes (no HCl) have optimum pH about 7.5

ENZYMES ARE PROTEINS WHICH ACT AS CATALYSTS IN LIVING ORGANISMS

Substrate molecules fit exactly onto an ACTIVE SITE on the enzyme

Substrate molecules react together to form a **product** which leaves the active site

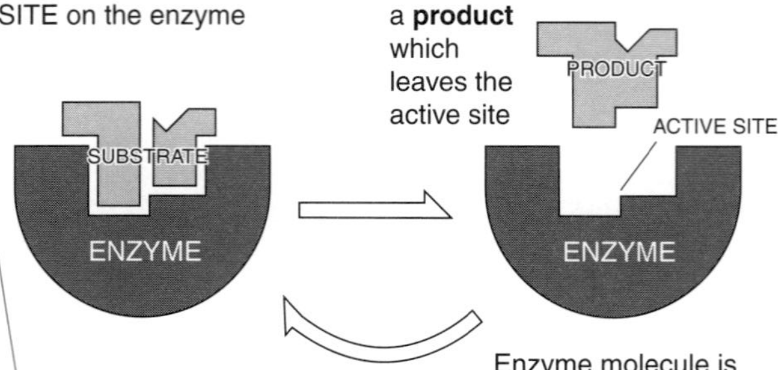

Enzyme molecule is now free to bind to more substrate molecules

The **shape** of the active site enables the enzyme to 'recognise' its substrate **in a very specific way**. Any factor which alters the enzyme's shape will affect its activity.

Inhibitors and activators: these are molecules which may
- **inhibit** by blocking the active site e.g. cyanide poisons by blocking enzymes in respiration
- **activate** by helping the active site to achieve its correct shape e.g. chloride ions in saliva activate the starch digesting enzyme salivary amylase

Enzymes may be intracellular or extracellular

these are both made **and** have their action inside cells ('intra' means 'inside')

e.g. photosynthetic enzymes inside chloroplasts
respiratory enzymes inside mitochondria.

these are **made** inside cells but **have their action** outside the cell ('extra' means 'outside')

e.g. digestive systems in the human gut
enzymes released by saprotrophic fungi and bacteria

THE SPECIFICITY AND CATALYTIC ACTIVITY OF ENZYMES MAKES THEM VERY USEFUL TO HUMANS

MEDICINE
- **Streptokinase** limits damage caused by heart attacks by dissolving blood clots
- **Urease** breaks down urea in dialysis fluid from kidney dialysis machines, allowing the fluid to be reused
- **Lactase** removes lactose from milk, thus making it safe for lactose-intolerant people

COMMERCIAL
- **Proteases** help to soften leather for the garment industry
- **Lipase** removes stains from clothing - component of 'biological' washing powders
- **Amylase** converts starch to sugars used in production of syrups, e.g. in fruit pies

PHARMACEUTICAL
- **Proteases** remove protein stains from false teeth
- **Catalase** removes hydrogen peroxide used to sterilise contact lenses

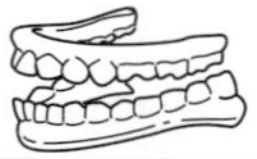

ANALYTICAL
- **Glucose oxidase** used in biosensors detects glucose levels in blood and urine
- **Carbonic anhydrase** detects levels of insecticides in water

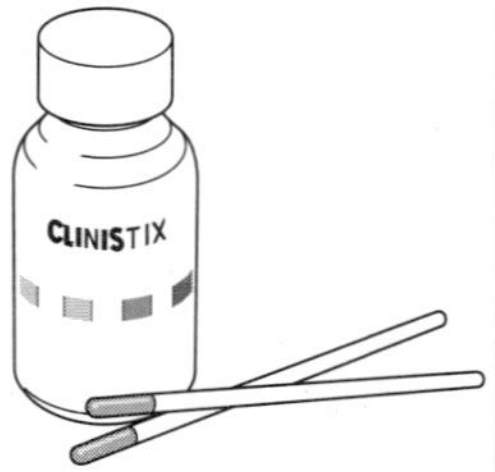

GENETIC ENGINEERING
- **Restriction enzymes** are used to cut out specific genes, and to open up bacterial plasmids
- **Ligases** are used to 'stitch' human genes into bacterial plasmids

Other, related topics can be found on pages 11, 46.

An ideal human diet contains fat, protein, carbohydrate, vitamins, minerals, water and fibre **in the correct proportions**.

An adequate diet provides sufficient **energy** for the performance of metabolic work, although the 'energy' could be in any form.

A balanced diet provides all dietary requirements **in the correct proportions**. Ideally this would be $\frac{1}{7}$ **fat**, $\frac{1}{7}$ **protein** and $\frac{5}{7}$ **carbohydrate**.

In conditions of **under**nutrition the first concern is usually provision of an **adequate diet**, but to avoid symptoms of **mal**nutrition a **balanced diet** must be provided.

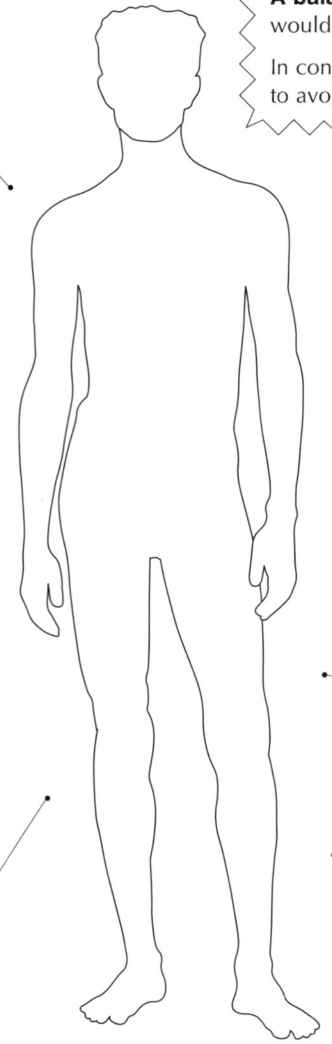

CARBOHYDRATES

Principally as an energy source. **Respiratory substrate** oxidised to release **energy** for active transport, synthesis of macromolecules, cell division and muscle contraction.

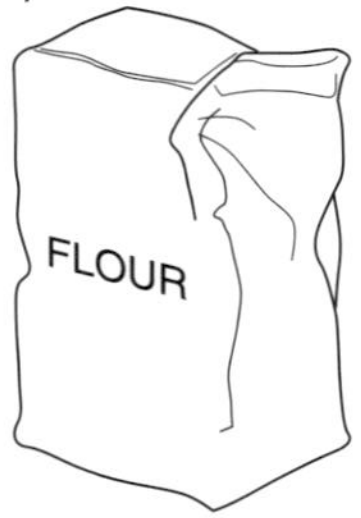

Common sources: rice, potatoes, wheat and other cereal grains i.e. as **starch**. and as refined sugar, **sucrose** in food sweetenings and preservatives.

Digested in mouth and small intestine and absorbed as **glucose**.

LIPIDS

An energy source (they are highly reduced and therefore can be oxidised to release **energy**). Also important in **cell membranes** and as a component of **steroid hormones**.

Common sources: Meat and animal foods are rich in **saturated fats** and **cholesterol**, plant sources such as sunflower and soya are rich in **unsaturated fats**.

Digested in the small intestine and absorbed as **fatty acids and glycerol**.

VITAMINS

VITAMINS have no common structure or function, but are essential in small amounts to use other dietary components efficiently. Their absence results in **deficiency diseases**.

e.g.	*Source*	*Deficiency causes*
vitamin C (water soluble)	Citrus fruits	**Scurvy** - bleeding gums, slow healing of wounds
vitamin D (fat soluble)	Cod liver oil, margarine	**Rickets** - misshapen and poorly growing bones

FIBRE

FIBRE (originally known as **roughage**) is mainly cellulose from plant cell walls and is common in fresh vegetables and cereals. It **may** provide some energy but mainly serves to aid faeces formation and prevent constipation.

PROTEINS

Are **building blocks** for growth and repair of many body tissues (e.g. myosin in muscle, collagen in connective tissues), as **enzymes**, as **transport systems** (e.g. haemoglobin), as **hormones** (e.g. insulin) and as **antibodies**.

Common source: meat, fish, eggs and legumes/pulses. Must contain eight **essential amino acids** since humans are not able to synthesise them. (Animal sources generally contain more of the essential amino acids than vegetable sources.) Digested in stomach and absorbed as **amino acids**.

Deficiency of protein causes poor growth - in extreme cases (in developing countries) may cause **marasmus** or **kwashiorkor**.

WATER

WATER is required as a solvent, a transport medium, a substrate in digestive reactions and for lubrication (e.g. in tears). A human requires 2–3dm^3 of water daily - most commonly from drinks and liquid foods.

MINERALS

MINERALS have a range of **specific** roles, and absence may cause **deficiency diseases**

e.g.	*Source*	*Deficiency causes*
Calcium (Ca^{2+})	Dairy products	Poor growth of bones and teeth
Iron (Fe^{2+})	Red meat, spinach	Anaemia - poor oxygen transport in blood

They are usually ingested with other foods, but supplements may be necessary (e.g. iron tablets are sometimes needed following menstruation).

Malnutrition

Malnutrition in the developed world usually means **"bad" nutrition** - the diet is usually more than adequate but may not be balanced.

Diets rich in refined sugar may cause **tooth decay**. Bacteria in the mouth convert sugar to acid which damages tooth enamel.

Diets high in saturated fats and **cholesterol** may cause **heart disease** - fatty compounds block the arteries which supply glucose and oxygen to heart muscle.

Diets high in salt may cause blood pressure to rise which in turn may cause **disorders of the circulation**.

Diets low in dietary fibre (roughage) may cause **constipation** and **cancer of the colon** - gut muscles become inefficient and faeces remains in the gut for too long.

Diets too high in energy content may cause **obesity** - excess 'energy' can be stored in unlimited amounts of fat. The extra mass and volume of body tissue cause

- joint damage
- increased risk of heart and circulatory diseases
- increased risk of diabetes
- infections of the lungs as breathing becomes difficult

Anorexia is an eating disorder in which food intake does not match the body's demands. It is often caused by peer group or 'fashion' pressures and has effects similar to marasmus

- muscles waste as they are used as a source of energy
- growth and sexual development are limited
- metabolic rate and blood pressure decrease
- depression and anxiety increase.

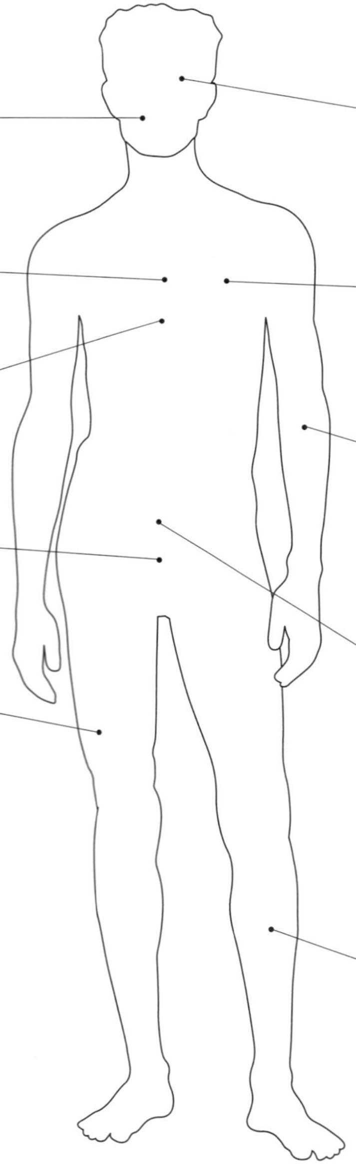

Malnutrition in developing countries usually means **undernutrition** - the diet may be neither balanced nor adequate.

Vitamin A deficiency causes **night blindness** (poor production of visual pigment) and **xerophthalmia** (hardening and flaking of the cornea).

Iron deficiency causes **anaemia**. Iron is required for haemoglobin synthesis so insufficient dietary iron causes oxygen transport by red blood cells to be limited and energy release by aerobic respiration is reduced.

Vitamin C deficiency causes **scurvy**. The production of some structural tissues is inhibited so that teeth become loose, skin becomes flaky and wounds bleed freely. Very low levels of vitamin C may also make the immune system less efficient so that the body becomes more prone to infection.

Protein deficiency may cause **kwashiorkor** - mental and physical development is slowed down. The swollen abdomen which occurs in such cases is due to insufficient blood protein to be able to reclaim water from the tissues and to an enlarged liver. If there is also a shortage of energy foods **marasmus** may occur - this is a general wasting of all body tissues and the body becomes very thin and wrinkled.

Deficiency of calcium or **vitamin D** may cause poor development of bones - this is seen as **rickets** in children (characterised by 'bow legs') and **osteoporosis** in adults (often leading to easy fracturing and poor healing of bones).

Digestion, absorption and processing of foods

Mouth: food is chewed using teeth and tongue to mix it with saliva and prepare it for swallowing.

STARCH —SALIVARY AMYLASE→ SUGARS*

Oesophagus: propels swallowed food to stomach, Has muscular walls which 'push' food with a wave of contraction - **peristalsis.**

Liver: produces bile.

Gall bladder: stores bile.

Stomach: muscular bag which stores and mixes food. Acidic juices ideal for enzyme activity.

PROTEINS —PEPSIN→ AMINO ACIDS*

Pancreas: produces alkaline juice (to neutralise stomach acid) which is rich in enzymes.

Small intestine: juices from gall bladder and pancreas complete digestion

STARCH —PANCREATIC AMYLASE→ SUGARS*

FATS —BILE→ FAT DROPLETS

FAT DROPLETS —LIPASE→ FATTY ACIDS* and GLYCEROL

* Digested foods are absorbed into the blood

Large intestine: absorbs water from gut contents.

Rectum: undigested foods are stored as faeces before expulsion from the anus.

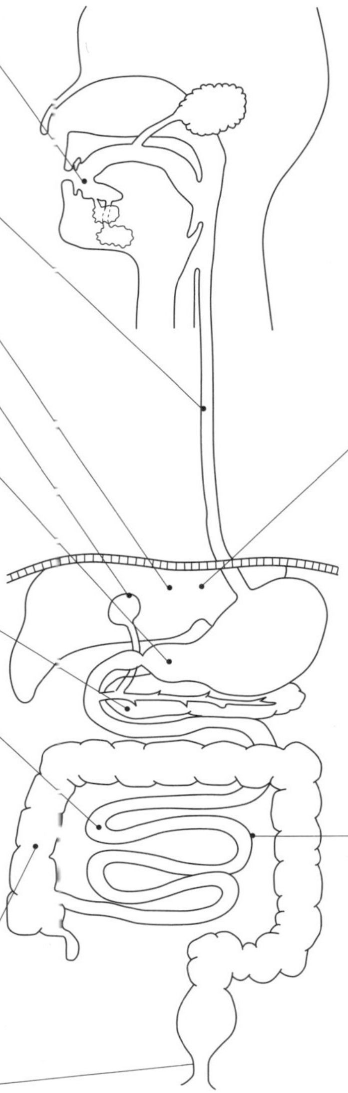

THE LIVER PROCESSES THE ABSORBED FOODS

Many cells perform a wide range of functions in processing the absorbed foods.

GLUCOSE ⇄ (GLUCAGON / INSULIN) GLYCOGEN STORES
GLUCOSE → ENERGY via RESPIRATION
GLUCOSE → to OTHER TISSUE via the circulation

AMINO ACIDS → SYNTHESIS OF PLASMA PROTEINS e.g. fibrinogen
AMINO ACIDS → EXCESS are DEAMINATED → UREA for excretion
AMINO ACIDS → to OTHER TISSUES via the circulation

FATTY ACIDS → FAT STORES
FATTY ACIDS → FATS for CELL MEMBRANES
FATTY ACIDS → ENERGY via RESPIRATION

All of this metabolic work generates heat - very important for endotherms as they use food energy to keep a constant body temperature.

THE SMALL INTESTINE IS ADAPTED FOR ABSORPTION

CROSS SECTION OF SMALL INTESTINE

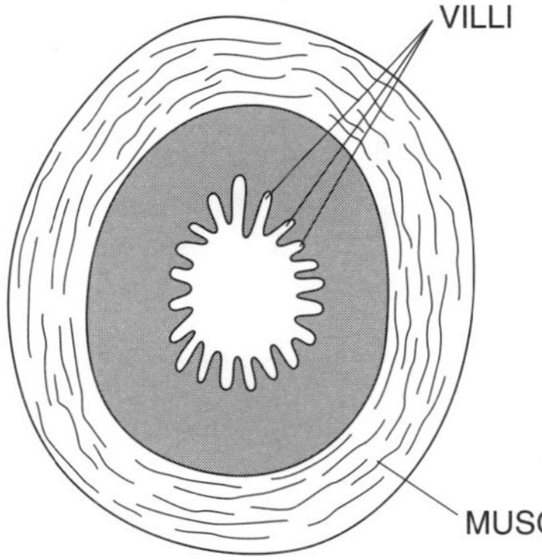

A SINGLE VILLUS

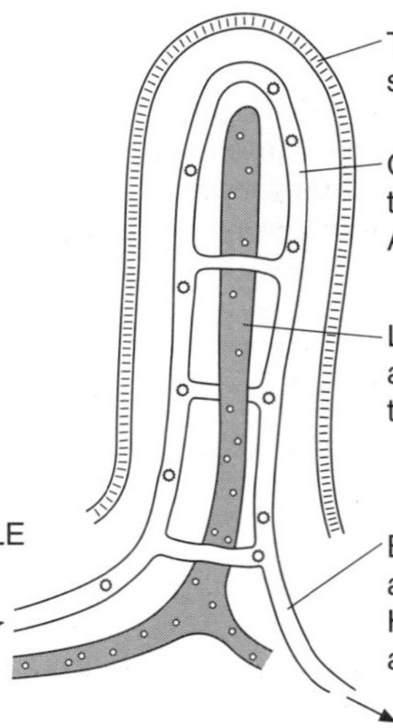

Note
- muscle layers for peristalsis
- villi to increase surface area for absorption

Semi-lunar (watch-pocket) valves prevent backflow of blood from arteries as the pressure in the ventricles falls.

THE HEART is the pump which applies pressure to force the blood around the circulation.

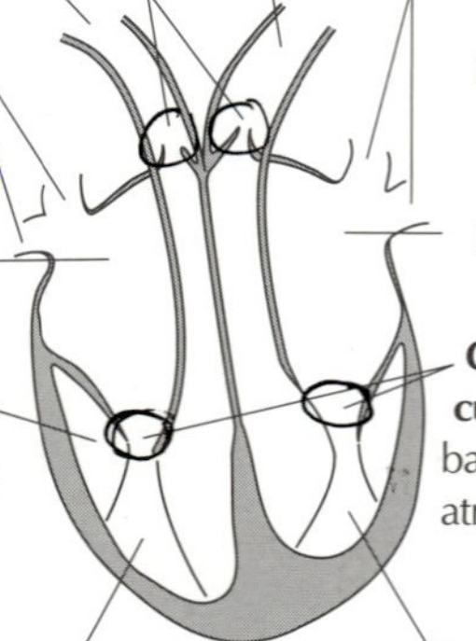

Pulmonary artery

Aorta

Vena cava

Pulmonary veins

Right atrium: wall contains the pacemaker of the heart

Left atrium: walls of atria are thin since need only push blood down to the ventricles

Heart 'strings' prevent valves turning inside out as pressure rises in ventricles.

Cuspid/atrioventricular valves prevent backflow of blood to atria as pressure rises in the ventricles

Right ventricle: less muscular than left ventricle since only pumps blood to lungs

Left ventricle: most muscular part of heart since must pump blood around body

HEART DISEASE affects 1 in 4 people between the ages of 30 and 60. High risk factors include

- Cigarette smoking
- High blood cholesterol
- Obesity
- Lack of regular exercise

A 'heart attack' involves death of an area of heart tissue following an interrupted blood supply - usually when the **coronary artery** (which delivers blood to heart muscle itself) is narrowed by fatty deposits.

Angina (severe chest pain following physical exertion) **may** be a forerunner to a full heart attack.

Mammalian circulation

Mammalian circulation is referred to as a **double** circulation since blood passes **twice** through the heart for each complete circuit of the body.

Pulmonary artery: unlike other arteries this vessel carries **deoxygenated blood** which also has a **high CO_2 concentration**.

Vena cava: the main vein of the body returns deoxygenated blood at low pressure from organs and tissues to the heart.

Renal vein: blood with **reduced urea concentration** is returned to the circulation.

Hepatic vein: blood with a regulated, optimum concentration of food substances is delivered to the circulation.

Veins have thin, non-muscular walls…

…but contain **semi-lunar valves** to prevent backflow of low pressure blood.

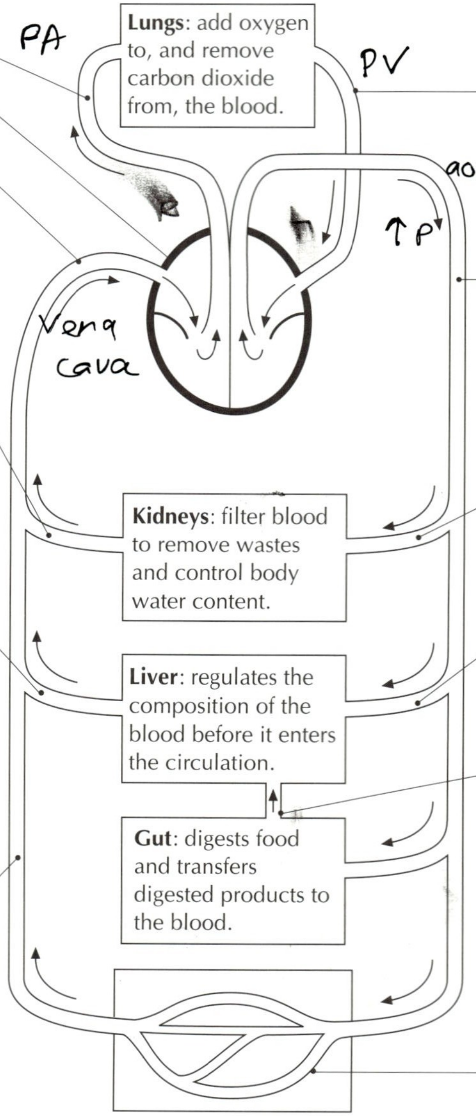

Pulmonary vein: unlike other veins this vessel carries **oxygenated blood** - it has the highest oxygen/lowest carbon dioxide concentration in the circulation.

Aorta: oxygenated blood, at high pressure, is delivered by this artery to the organs and tissues of the body. All arteries have a thick, muscular, elastic wall but contain no valves. The **pulse** can be felt where an artery crosses a bone.

Renal artery: oxygenated blood with a high concentration of urea is delivered to the kidneys.

Hepatic artery: carries oxygenated blood to the liver.

Hepatic portal vein: carries blood containing variable amounts of the absorbed products of digestion from gut to the liver.

Capillaries are present in all organs/tissues and are the site of exchanges of materials between blood and tissue fluid

- very high surface area
- walls only one cell thick

make sure that diffusion is rapid enough to satisfy the needs of the tissues.

Blood structure and function

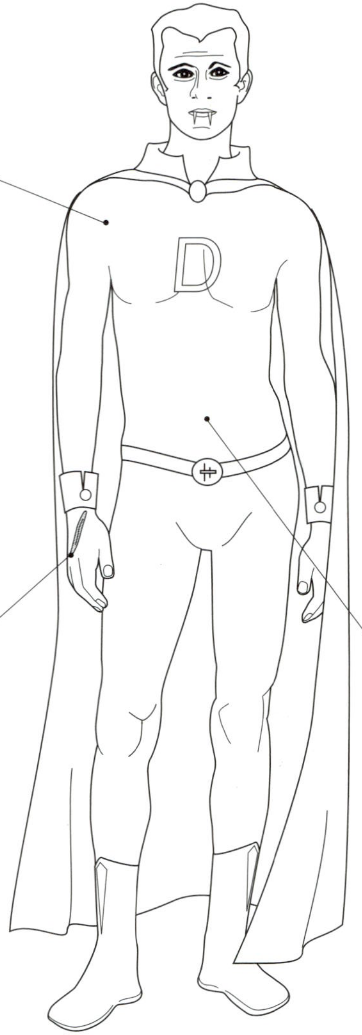

REGULATORY FUNCTIONS - HOMEOSTASIS
Blood solutes affect the water potential of the blood, and thus the water potential gradient between the blood and the tissue fluid. The size of this water potential gradient is largely due to sodium ions and plasma proteins and **regulates the movement of water** between blood and tissues.

Water plays a part in the **distribution of heat** between heat producing areas such as the liver and areas of heat loss such as the skin.

Too much blood alcohol can cause water to leave brain cells causing the pain and sensation of thirst called a **hangover!**

Blood also helps to maintain **optimum pH** in the tissues.

PROTECTIVE FUNCTIONS
Platelets, plasma proteins (e.g. fibrinogen) and many other plasma factors (e.g. Ca^{2+} ions) protect against **blood loss** and the **entry of pathogens** by the clotting mechanism.

White blood cells protect against **disease-causing organisms**:
phagocytes engulf them
lymphocytes produce and secrete specific antibodies against them.

Dracula may have consumed blood because he suffered from **porphyria**, a condition characterised by an inability to synthesise some compounds including the haem group of haemoglobin, a marked sensitivity to daylight and abnormal development of the teeth!

COMPONENTS OF BLOOD: if a blood sample is allowed to settle, or spun for a few minutes in a centrifuge, it separates into two distinct layers

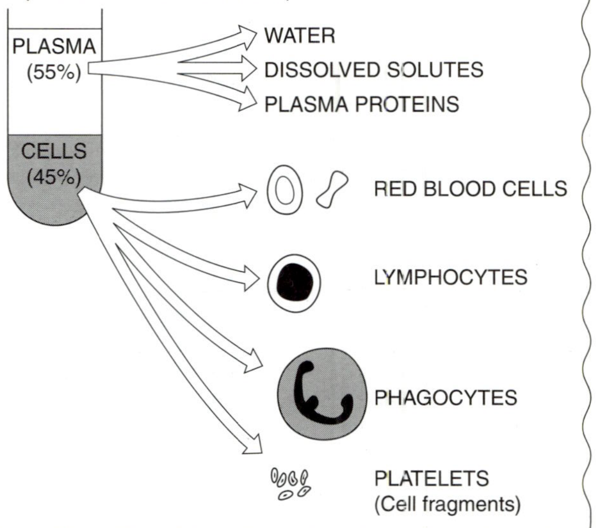

TRANSPORT FUNCTIONS
Soluble products of digestion/absorption (such as glucose, amino acids, fatty acids, vitamins and minerals) from the gut to the liver and then to the general circulation.

Waste products of metabolism (such as urea, creatinine and lactate) from sites of production to sites of removal, such as the liver and kidney.

Respiratory gases (oxygen and carbon dioxide) from their sites of uptake or production to their site of use or removal. Oxygen transport is more closely associated with red blood cells, and carbon dioxide transport with the plasma.

Hormones (such as insulin) from their sites of production in the glands to the target organs where they exert their effects.

Disease may have a number of causes

During **disease** the body's activities deviate from their normal levels by an amount which is more than can be counteracted by the usual homeostatic mechanisms.

Typical symptoms of disease include
- **sweating/fever** due to resetting of body's thermostat
- **vomiting/diarrhoea** due to attempt to 'clear' gut of irritants
- **pain** due to release of toxins by pathogens.

INFECTIOUS DISEASES ARE CAUSED BY PATHOGENS - these may spread in a number of ways:

in **droplets in the air**
e.g. Influenza virus

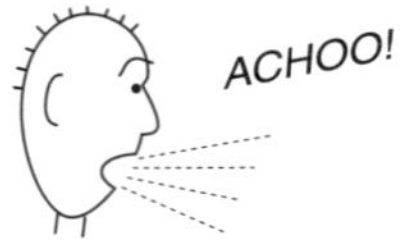

in **contaminated food**
e.g. *Salmonella* bacteria

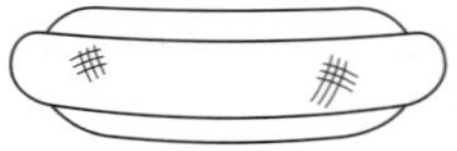

in **infected water**
e.g. *Cholera* bacteria

by **direct contact** (**contagious**)
e.g. Athlete's foot-fungus

via **body fluids**
e.g. Hepatitis B virus

by **animal vectors**
e.g. *Plasmodium* protozoan (causes malaria) via *Anopheles* mosquito

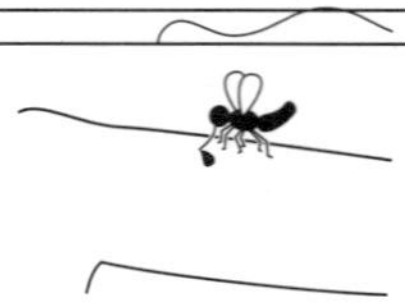

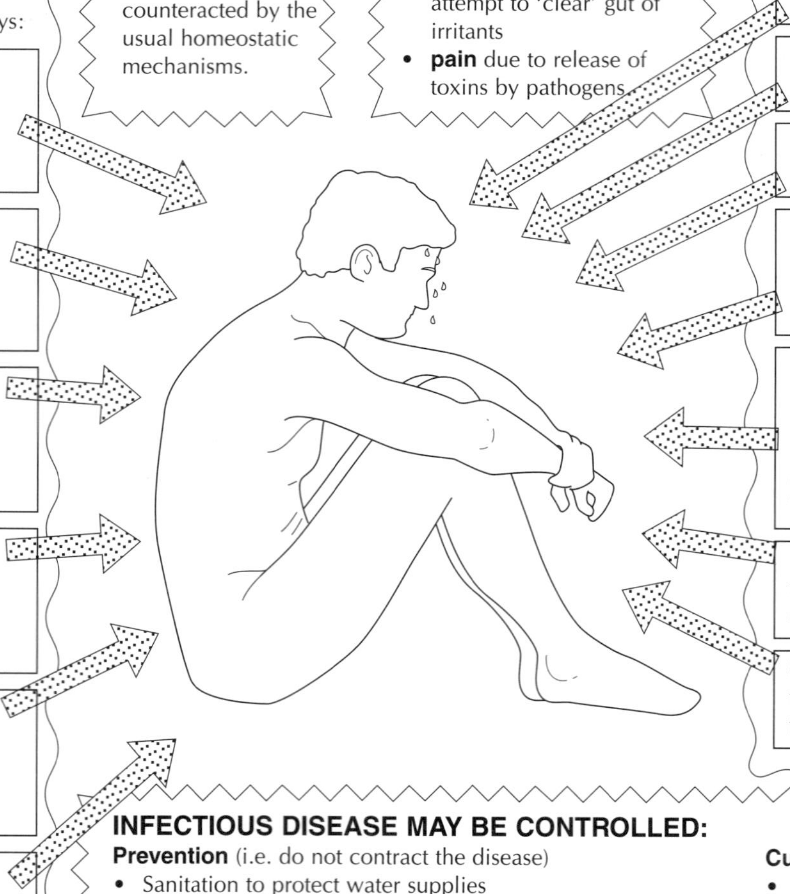

SOME DISEASES ARE NOT INFECTIOUS. They are **not** caused by pathogens. They may be

Degenerative: organs and tissues may work less well as they age. This is thought to be due to changes to body chemicals caused by **free radicals** such as the peroxide ion e.g. heart attacks, cataracts, hardening of the arteries.

Deficiency: poor diet may deprive the body of some essential substance e.g. scurvy (lack of vitamin C).

Allergy: sensitivity to some environmental antigen e.g. hayfever (pollen is the antigen).

Environmental: some factor in the environment may trigger a dangerous or abnormal bodily reaction e.g. overexposure to ultraviolet radiation may cause abnormal cell division leading to skin cancer.

Inherited/metabolic: some failure in the body's normal set of chemical reactions
e.g. sickle cell anaemia (abnormal haemoglobin)
cystic fibrosis (production of thick mucus)
diabetes (failure to produce enough insulin)
These conditions are due to alterations in the genes.

Psychological/mental: some changes in the working of the brain may lead to abnormal behaviour e.g. schizophrenia, depression.

Self-induced: some abuse of the body may affect its function e.g. lung cancer (cigarette smoking), cirrhosis of the liver (alcohol abuse).

INFECTIOUS DISEASE MAY BE CONTROLLED:

Prevention (i.e. do not contract the disease)
- Sanitation to protect water supplies
- Effective farming, e.g. of chickens, to prevent food poisoning
- Hygiene to prevent spread of contagious diseases. Use of **antiseptics** to kill pathogens before they enter the body.
- Destruction of vectors such as mosquitoes with insecticides
- Vaccination to 'prime' the immune system.

Cure (i.e. treat the patient with the disease)
- **Antibiotics** to destroy bacteria
- **Passive immunity** to provide ready-made antibodies
- **Specific drugs** to kill abnormal cells
- **Pain relievers** to limit discomfort/speed recovery.
- **Radiotherapy** to prevent division of abnormal cells.

Food preservation
may prevent disease

Milk pasteurisation is a valuable technique since
- it destroys most bacterial cells
- it does not alter the flavour of the milk.

The **flash** process involves
- heating food rapidly to 72°C (**not** boiling)
- hold at this temperature for 15 seconds
- cool rapidly (and keep **refrigerated**).

Saprotrophs e.g. fungi and bacteria are responsible for the decomposition of food. This causes two problems:

- as the food decays **less is available for humans**, either because the food has been 'consumed' or because its texture or appearance become less attractive.
- the saprotroph may release compounds onto the food - these may be **toxins** (poisons) which can cause discomfort or illness.

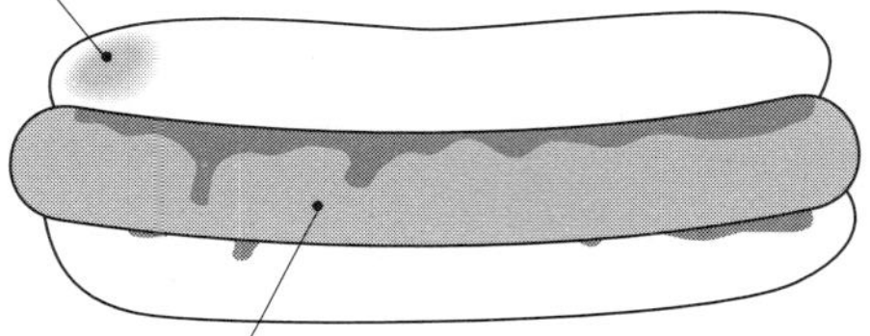

TECHNIQUES OF FOOD PRESERVATION

Sterilisation
- Food heated to high temperature, then sealed in an air-tight container
- destroys bacterial cells, but not their spores
- may alter the taste of the food.

Refrigeration
- keep food between 0–4°C
- does not **kill** bacteria, but slows their reproduction - thus the food is only preserved for 48–72 hours.

Irradiation
- food is exposed to γ-radiation
- microorganisms are destroyed, but enzymes are not. Thus irradiated food is not affected in texture or in ripening time, and this makes this technique attractive to customers.

Irradiated food does not contain microorganisms but as in other preservation techniques, **toxins are not destroyed** - thus irradiated food may still be dangerous.

Chemical methods
- use preservative substances to kill bacteria or slow down their reproduction
- examples include sulphur dioxide and nitrate compounds. Such substances are **additives** and can be identified by having an 'E number' e.g. E270.

***Salmonella* can cause food poisoning and can occur in undercooked meat or infected eggs.**

Intensive rearing of chickens makes it easy for *Salmonella* bacteria to spread from chicken gut → faeces → floor → chicken carcasses.

68°C cooking to this temperature kills *Salmonella* in meat.

Salmonella in gut multiplies under ideal conditions of temperature/food availability and releases a **toxin**. This inflames the gut lining → fever, pain, vomiting and diarrhoea → dehydration, and even death.

Treatment: antibiotics are **not** very effective because they do not easily reach the gut lining.

Replace body fluids with a mixture of water, glucose and salts.

WATER MUST BE STERILISED
- A water supply is necessary for **drinking**, **washing** and **preparation of food**.
- Important water-borne diseases include:
  Typhoid (transmitted in faeces of infected people)
  Cholera (water and shellfish may be contaminated)
  Dysentery (can be caused by bacteria or by an amoeba which is transmitted in infected faeces)
- Water is purified by **sewage treatment** and by the use of **chlorine**, which directly kills the infecting organisms.

Natural defence systems of the body

prevent infection and disease.

PATHOGENS CAUSE DISEASE. Many organisms may colonise the body of a human - the body is **warm**, **moist** and a **good food source**. These organisms may compete with human cells for nutrients or may produce by-products which are poisonous to human cells. This will affect the normal function of the cells i.e. it will cause 'disease'. There are three major groups of pathogens:

Type of pathogen	Disease	Symptoms
VIRUS 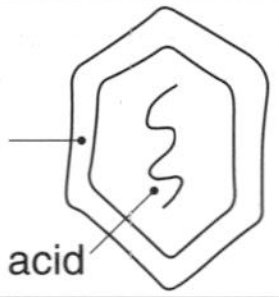	Influenza	Fever (raised body temperature). Aching joints. Breathing problems. **Control by pain relief/ rest/drinking fluids.**
BACTERIUM 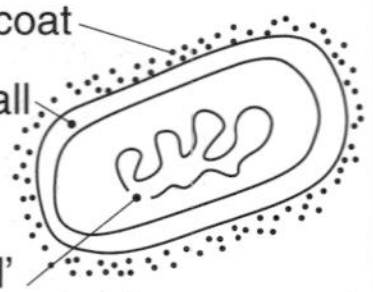	Gonorrhoea	Painful to urinate - yellow discharge from penis or vagina. In the long-term may cause blockage of sperm ducts/oviducts leading to sterility. **Control with antibiotics.**
FUNGUS 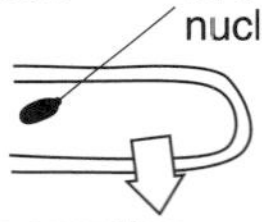	Athlete's foot	Irritation to moist areas of skin (e.g. between toes). 'Cracked' skin may become infected. **Control with fungicide/drying powders.**

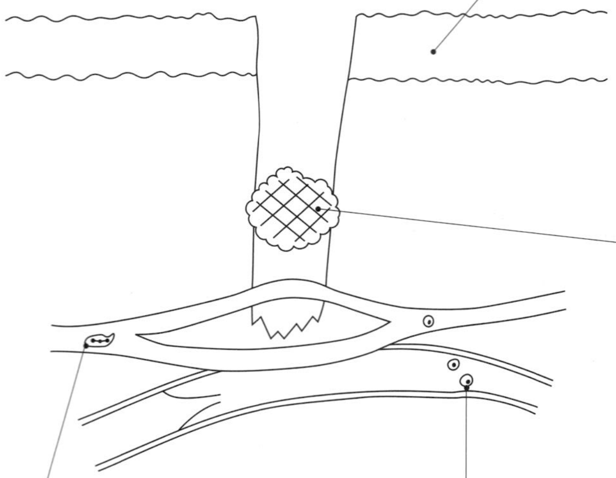

SKIN IS THE FIRST BARRIER: the outer layer of the skin, the **epidermis**, is waxy and impermeable to water and to pathogens (although many microorganisms can live on its surface). Where there are natural 'gaps' in the skin there may be protective secretions to prevent entry of pathogens, e.g.

Orifice	Function	Protected by
Mouth	Entry of food	Hydrochloric acid in stomach
Eyes	Entry of light	Lysozyme in tears
Ears	Entry of sound	Bacteriocidal ('bacteria-killing') wax

BLOOD CLOTTING 'PLUGS' WOUNDS is a natural defence, largely due to **blood proteins** and **platelets**. It is able to block any unnatural gaps in the skin.

- prevents excessive blood loss
- prevents entry of pathogens

DAMAGED PLATELETS / TORN CAPILLARIES → ENZYMES

Inactive blood proteins e.g. FIBRINOGEN → (If CALCIUM IONS present) → Active blood proteins e.g. FIBRIN → FIBRES

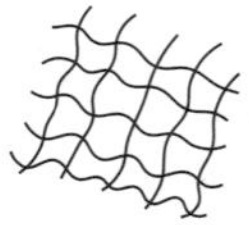

Mesh can trap red blood cells which dry out in form of a scab

Absence of any of these blood proteins causes inefficient clotting → severe bleeding: **haemophilia**.

PHAGOCYTES DESTROY PATHOGENS BY INGESTING THEM
Phagocytes are large white blood cells. They are 'attracted to' wounds or sites of infection by chemical messages. They leave the blood vessels and destroy any pathogens which they recognise.

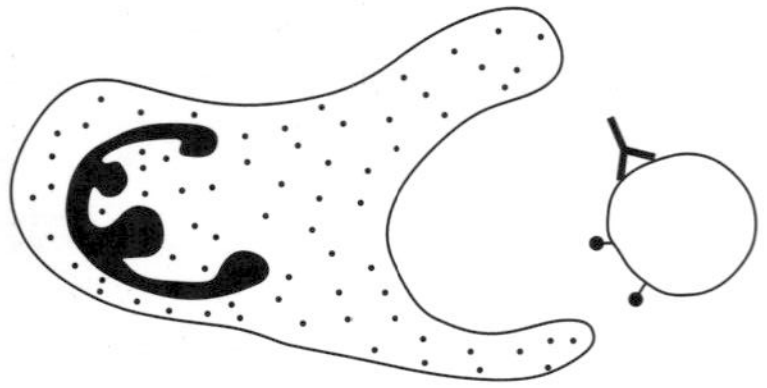

A pathogen is recognised by its surface proteins or by antibodies which 'label' it as dangerous.

Pathogens are destroyed by digestive enzymes secreted into 'food sac'.

LYMPHOCYTES PRODUCE PROTECTIVE ANTIBODIES
Lymphocytes are white blood cells which are found both in the blood and in lymph nodes (swellings in the lymphatic system). They are stimulated by the presence of pathogens to manufacture and release special proteins called **antibodies** which can recognise, bind to, and help to destroy pathogens.

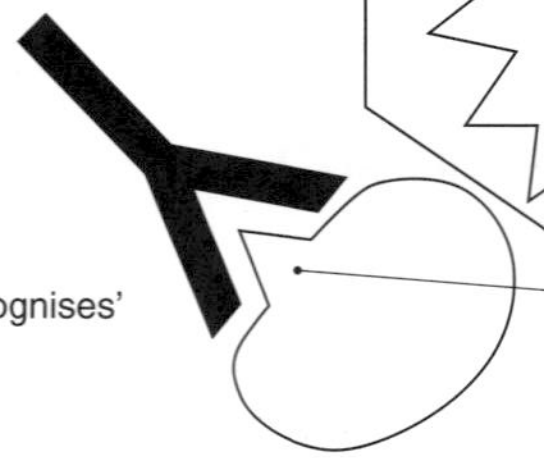

Antibodies are specific protein molecules.

THE IMMUNE RESPONSE MAY BE ENHANCED. Immunity is the body's use of antibodies to combat invasion by pathogens.

Immunity may be

- **ACTIVE: the individual is provoked to make his or her own antibodies**
  - NATURAL – Pathogen infects individual: Individual contracts disease but survives, makes antibodies and is now immune to further infection by the same pathogen e.g. immunity develops to different strains of the common cold.
  - ARTIFICIAL – Weakened pathogen (vaccine): Injection of vaccine does not cause disease but lymphocytes do produce antibodies – individual now immune to this pathogen e.g. vaccination against rubella virus (causes German measles) in babies.
- **PASSIVE: the individual is protected by a supply of pre-formed antibodies**
  - NATURAL: Mother's antibodies can cross the placenta and are delivered in breast milk: newborn child will be temporarily immune to pathogens for which mother produced antibodies.
  - ARTIFICIAL: Antibodies collected from blood of laboratory animal e.g. horse or rabbit. → INJECTION → Adult is now immune to disease which is too fast-acting for own immune system to deal with e.g. injection of anti-tetanus vaccine following a deep, dirty cut or wound. NB This offers only a TEMPORARY immunity.

VACCINE PRODUCTION

Use either

- Dead pathogens e.g. Whooping cough vaccine.
- Weakened pathogens e.g. Oral polio vaccine.
- Genetically engineered fragments – the proteins from the pathogens surface which are recognised by lymphocytes e.g. Hepatitis B viral coat protein

MONOCLONAL ANTIBODIES are specific protein molecules with important medical and industrial uses

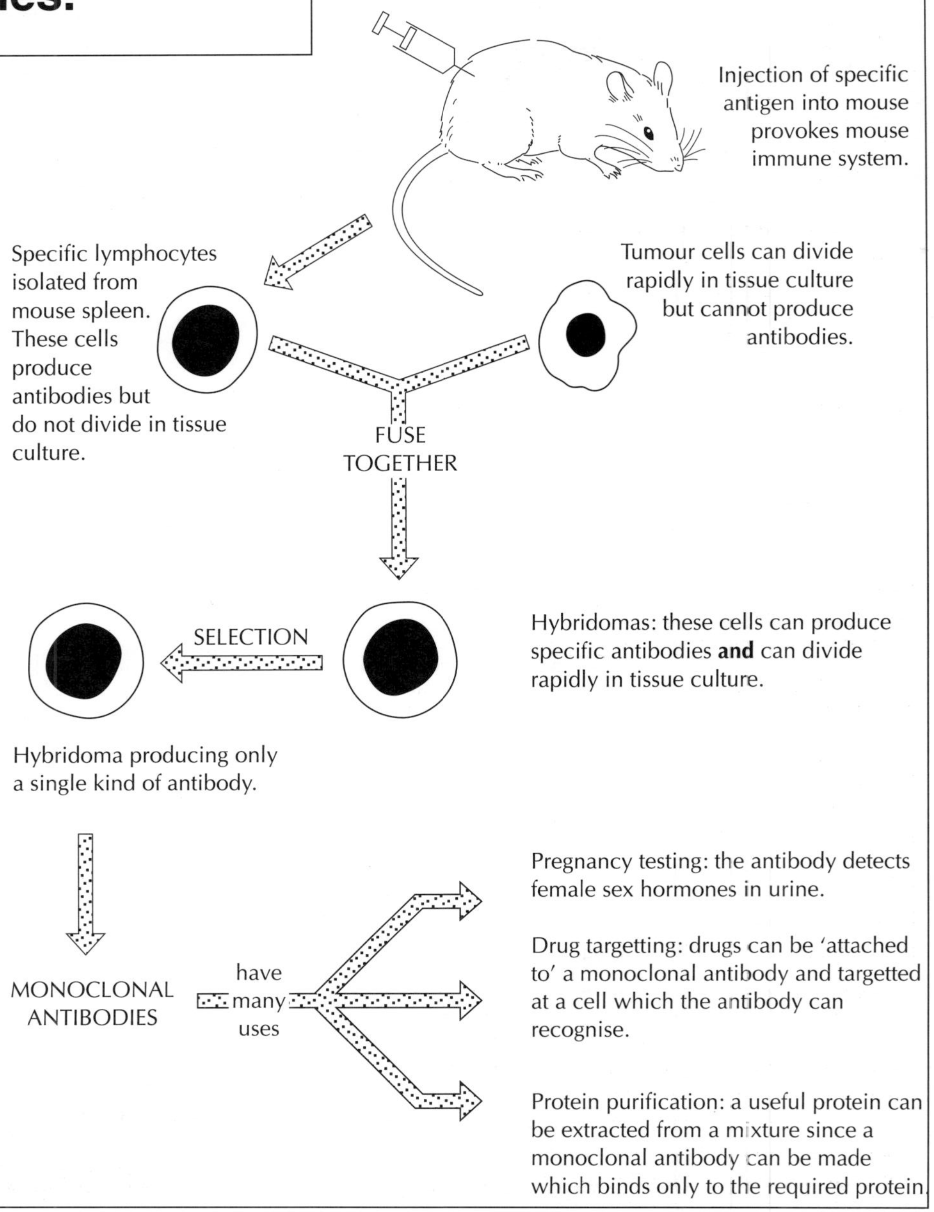

Abuse of drugs

SOLVENT ABUSE causes both short- and long-term damage.

- damages membranes of brain cells: **loss of memory, irrational behaviour**.
- damages membranes which line the nose: **more infections of breathing passages**.
- very high doses of, e.g. butane, cause constriction of airways: **death by asphyxiation**.

SMOKING OF TOBACCO has many harmful effects.

- inhibits cilia which normally sweep a stream of mucus up the bronchi to trap and remove dust, pollen and pathogens. Lining of bronchi become irritated and infected: **bronchitis**.
- carcinogens provoke uncontrolled cell division in tissues of airways - tumours develop, block airways and invade adjacent tissues: **lung cancer**.
- changes function of white blood cells in lungs which become overactive and 'digest' lung tissue: **emphysema**.
- carbon monoxide irreversibly binds to haemoglobin and reduces oxygen-carrying ability of blood - heart 'pumps' harder to compensate: **heart failure**.
- increases deposition of fats in the arteries thus reducing circulation - tissues die: **limb amputations**.
- residues to be excreted are stored as part of urine in the bladder. Cells divide uncontrollably: **bladder cancer**.

OVERUSE OF ANTIBIOTICS may clear the harmless bacteria and fungi from the gut, allowing possibly dangerous organisms to colonise the now-vacant 'habitat': **severe gut infections**, **diarrhoea**.

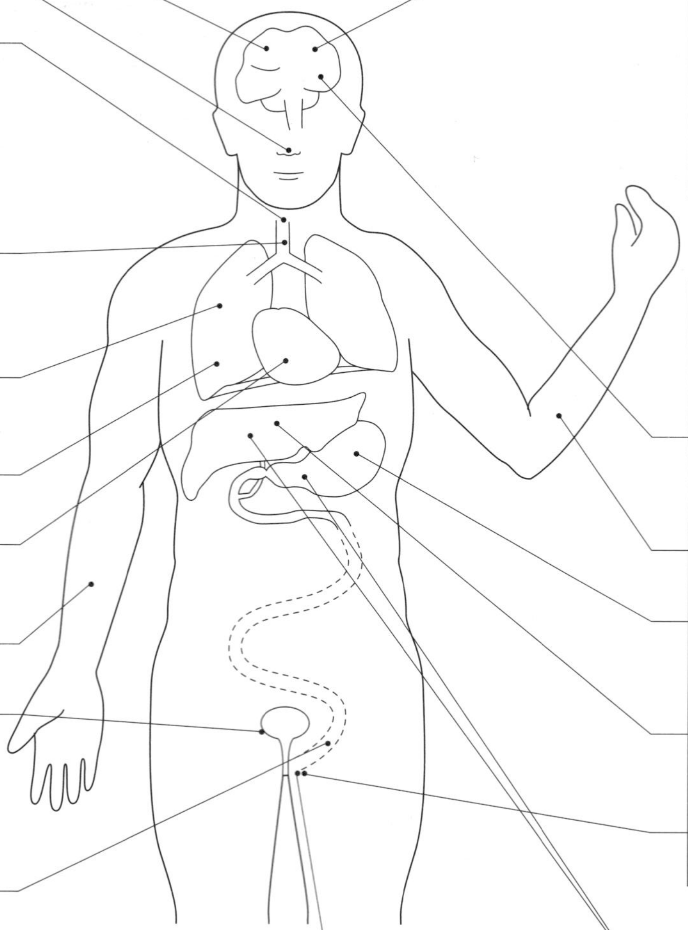

ABUSE OF STEROID HORMONES may raise blood levels of these compounds so that they feedback inhibit natural production of sex hormones: **sterility**, **loss of sexual desire**.

MANY DRUGS AFFECT BRAIN FUNCTION

Barbiturates (taken as sleeping tablets) can depress breathing centre: **death**.

L.S.D. (a **hallucinogen**): affects ability to 'sort out' messages leading to faulty 'connections': **hallucinations**

Heroin: causes euphoria (a 'high') by depressing input of 'pain/discomfort' messages: **poor appetite**, **loss of awareness of surroundings**.

Caffeine (taken as a stimulant)increases sensitivity of nerve endings: **hyperactivity**.

OVER CONSUMPTION OF ALCOHOL has many harmful effects, both long- and short-term.

- inhibition of nerve impulse conduction: **slowed reflexes**, **dangerous driving**, **death**.
- depression of visual centre of brain: **poor co-ordination**, **accidents**.
- depression of breathing centre of brain: **death**.
- dilation of blood vessels of skin: **heat loss**, **hypothermia**, **death**.
- depression of appetite: **malnutrition**.
- irritation of lining of stomach: **vomiting - inhalation of vomit may cause death**.
- conversion to harmful product in liver causes hardening of lining of liver passages: **cirrhosis** - eventual liver failure: **death**.
- changes in blood pressure: **reduction in sexual performance**.

OVERUSE OF PAINKILLERS MAY BE DANGEROUS.

- excess paracetamol is converted to a poison by the liver, causing liver failure: **death**.
- excess aspirin reduces blood clotting ability: **bleeding of stomach lining**.

Neurones (nerve cells) carry information through the nervous system.

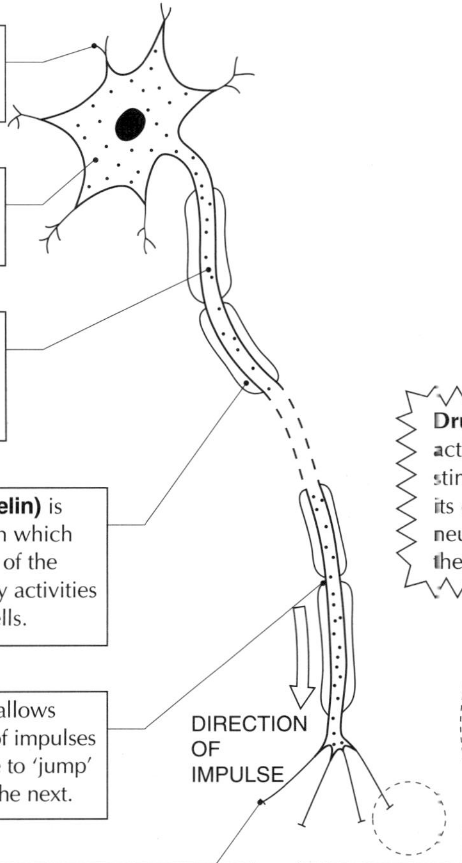

Dendrites 'collect' information from other cells.

Cell body controls the metabolism of the nerve cell.

Axon is extended to carry information over long distances.

Fatty sheath (myelin) is electrical insulation which prevents distortion of the nervous impulse by activities of neighbouring cells.

Node of Ranvier allows rapid conduction of impulses by forcing message to 'jump' from one node to the next.

End plate 'synapses' with another nerve cell, a muscle or a gland.

Drugs may interfere with synaptic activity e.g. Amphetamine ('Speed') stimulates nervous activity because its chemical structure is similar to the neurotransmitter and it mimics it in the synaptic gap.

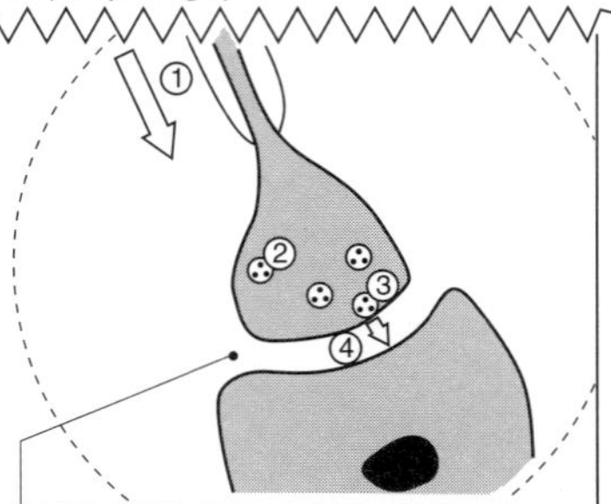

The synapse is the junction between the end plate of one nerve cell and the dendrite of the next.

1. An impulse arrives at the synapse
2. At the end plate are tiny sacs containing a chemical (**neurotransmitter)**
3. The chemical is released into the synaptic gap
4. The chemical diffuses across the gap and the impulse 'restarts' on the other side.

The reflex arc is the basic working unit of the nervous system and links **receptors** to **effectors** via the **integrators** of the central nervous system.

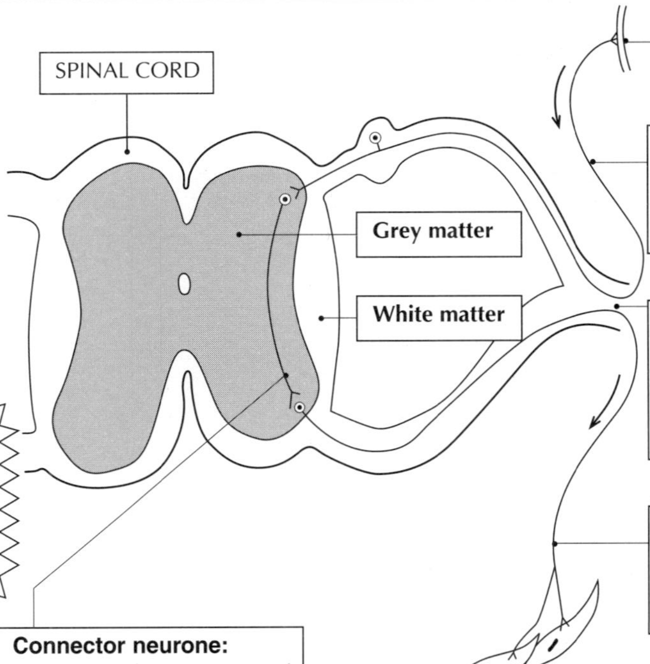

Receptor: a cell or organ which receives a stimulus and converts it into an electrical impulse.

Sensory neurone: carries impulse from the receptor to the central nervous system.

Spinal nerve: has both sensory and motor neurones, and leaves spinal column between adjacent vertebrae.

Motor neurone: carries impulse from the central nervous system to the effector.

Effector: muscle or gland which carries out an action (response) to 'deal with' the initial stimulus.

Connector neurone: carries impulse across spinal cord. This allows the action to be modified by impulses carried down the spinal cord from the brain.

REFLEX ACTIONS ALWAYS HAVE A POSITIVE SURVIVAL VALUE

e.g.

REFLEX	STIMULUS	RESPONSE	SURVIVAL VALUE
Pupil	Bright light falling on eye	Contraction of circular muscles of iris to reduce pupil size	Prevents 'bleaching' of retina which would lead to temporary blindness
Knee-jerk	Extension of patellar (knee) tendon e.g. as knee bends	Contraction of thigh muscle causing leg to straighten	Ensure that leg will support body when standing or, more especially, when walking

The endocrine system

produces and secretes chemical messengers called **hormones**.

1 Stimulus affects endocrine gland so that it releases a hormone (chemical messenger).

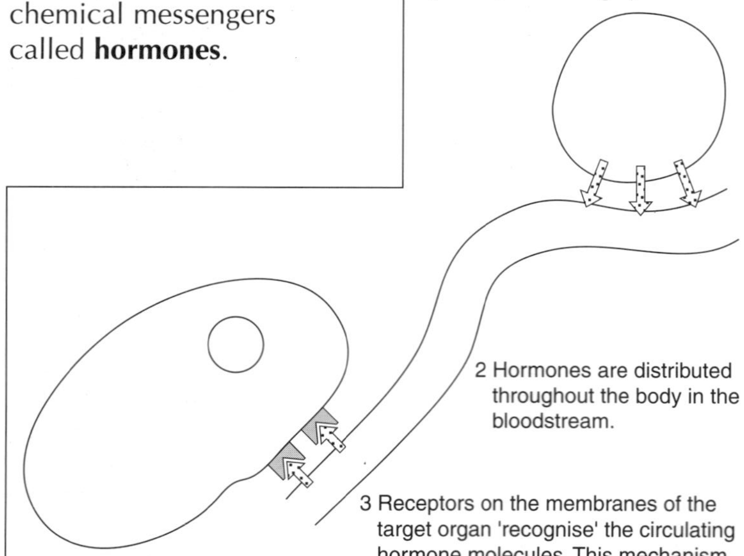

2 Hormones are distributed throughout the body in the bloodstream.

3 Receptors on the membranes of the target organ 'recognise' the circulating hormone molecules. This mechanism ensures that only the specific target organ can respond to the hormone.

4 Target organ brings about an appropriate response to deal with the original stimulus.

NERVOUS AND CHEMICAL CO-ORDINATION COMPARED

	NERVOUS	CHEMICAL (ENDOCRINE)
NATURE OF MESSAGE	Electrochemical impulses	Chemical compounds (hormones)
ROUTE OF TRANSMISSION	Specific nerve cells	General blood system
TYPE OF EFFECTS	Rapid, but usually short-term e.g. blinking	Usually slower, but generally longer-lasting e.g. growth

Note that the hormone **adrenaline** has effects which are rapid but short-term, and therefore partially mimics some effects of the nervous system.

SOME IMPORTANT APPLICATIONS OF HORMONES

- human growth hormone may be used to treat children who are growing very slowly
- insulin is used to control diabetes
- oestrogen and progesterone may be combined in the contraceptive pills to inhibit ovulation
- oestrogen may be used to promote milk production in cattle.

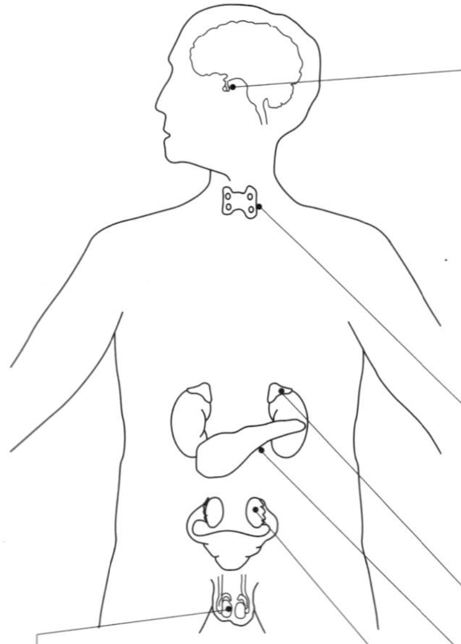

Pituitary gland secretes
- a number of hormones (trophins) which influence other endocrine organs
- human growth hormone
- anti diuretic hormone.

Thyroid gland secretes thyroxine which controls metabolic rate and is important in regulation of body temperature.

Pancreas secretes insulin (reduces blood glucose) and glucagon (raises blood glucose concentration).

Adrenal glands secrete **adrenaline** which is called the **fight** or **flight hormone** because it prepares the body for action.

- more glycogen is converted to glucose
- deeper, more rapid breathing
- faster heartbeat
- diversion of blood from gut to muscles.

The above all help to provide more glucose and more oxygen for the working muscles. There may also be
- dilation of the pupils
- raising of the hair.

Testes secrete testosterone, which
- controls the production and development of sperm
- regulates the development of the male secondary sexual characteristics

i.e. enlargement of sex organs
growth of facial hair
muscle enlargement
deepening of voice
sexual behaviour

Artificial steroids increase muscle bulk but inhibit natural testosterone production which can lead to sterility.

Ovaries secrete the female sex hormones.

Oestrogen
- controls the development of the female secondary sexual characteristics, including the pattern of laying down of fat which gives the feminine body shape
- controls the repair of the lining of the uterus
- helps to regulate the development of further follicles.

Progesterone
- prepares the lining of the uterus for implantation by increasing the thickness of the lining and number of blood vessels
- helps to inhibit the growth of further follicles.

The eye is a sense organ -

different tissues work together to perform one function...

... **retina** contains light sensitive cells
... **cornea** and **lens** focus the light onto the retina
... **sclera**, **choroid** and **iris** protect and support the eye.

Accommodation: is the process of producing a finely-focused image on the retina. It is carried out by the action of the **ciliary muscles** on the **lens**.

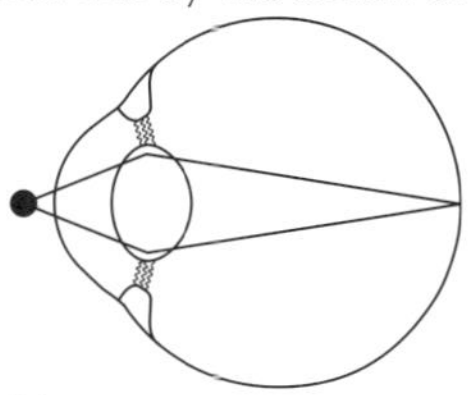

Close object:
light must be **greatly refracted** (bent)
ciliary muscles **contract**, pull eyeball inwards
ligaments **relax**
lens becomes **short and fat**.

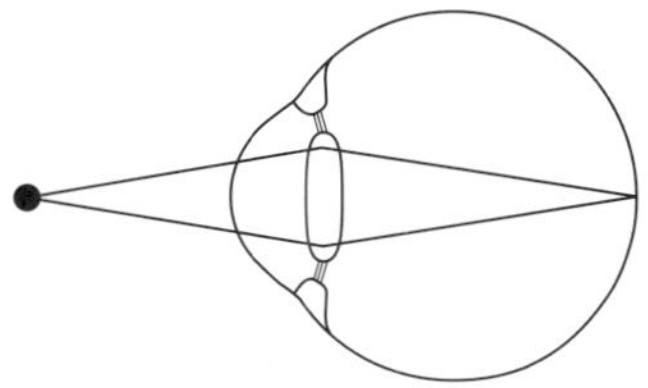

Distant object:
light can be **less refracted**
ciliary muscles **relax**, eyeball becomes spherical
ligaments **tauten**
lens is pulled **long and thin**.

Rods: provide **black-and-white** images but, because several may be 'wired' to a single sensory neurone in the optic nerve, they provide **great sensitivity** at low light intensity (night vision), but images lack detail.

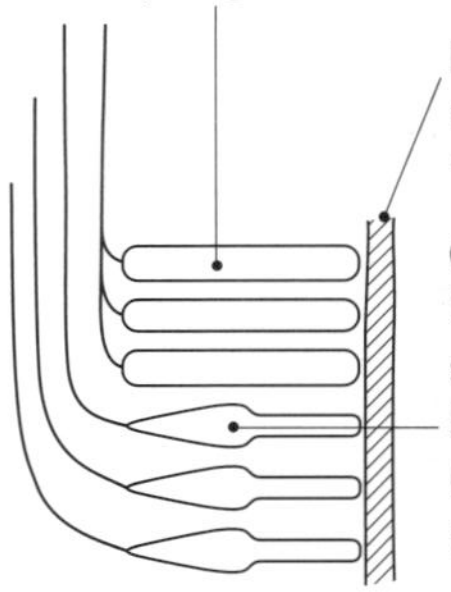

Layer of pigment: prevents internal reflection which might lead to 'multiple/blurred' images.

Cones: provide very **detailed** images, in **colour** (there are three types, sensitive to red, green and blue light) but **only under high light intensity** (their connections to the optic nerve make them rather insensitive).

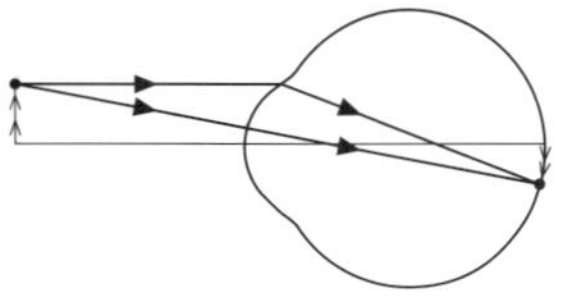

Cornea: a transparent layer which is responsible for most of the refraction (bending) of light rays from the object to form the inverted, and smaller image on the retina.

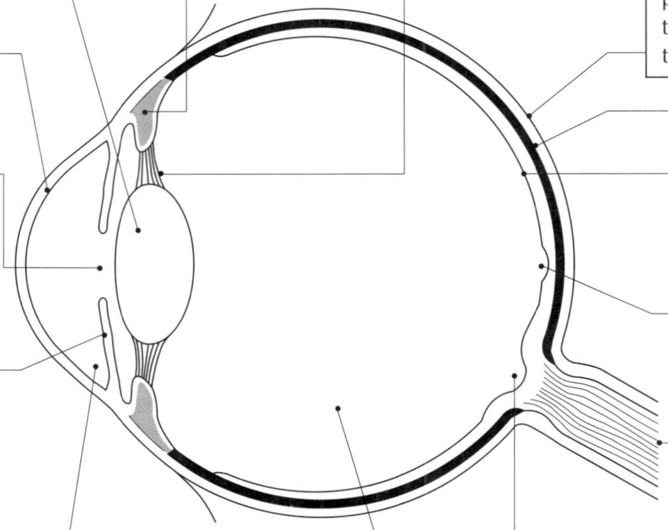

Lens

Ciliary muscle

Suspensory ligament

Sclera: the tough outer coat which protects the eye against damage and provides attachment for the muscles which move the eye in its socket.

Choroid: a darkly coloured layer which (a) limits internal reflections and (b) contains blood vessels which help to nourish the cells of the retina.

Pupil: the circular opening which lets light into the eye. Appears 'black' because the choroid is visible through it.

Retina: contains the light-sensitive cells, the rods and cones. Light arriving at this layer will produce an **inverted**, **smaller image**.

Yellow spot (Fovea): has the greatest density of cones and thus offers **maximum sharpness** but only works at full efficiency in **bright light**.

Optic nerve: composed of sensory neurones which carry nervous impulses to the **visual centre** at the rear of the brain.

Blind spot: corresponds to the exit point for the optic nerve. There are no light-sensitive cells here so that light falling on this region cannot be detected.

Iris: the 'coloured' part of the eye which may expand and contract to control the amount of light which enters the eye - this is a **reflex action**.

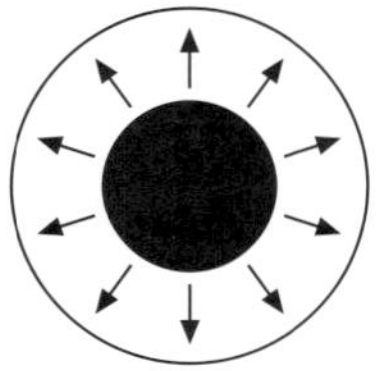

Low light intensity: radial muscles of iris contract and the pupil is opened wider - **more light may enter and reach retina**.

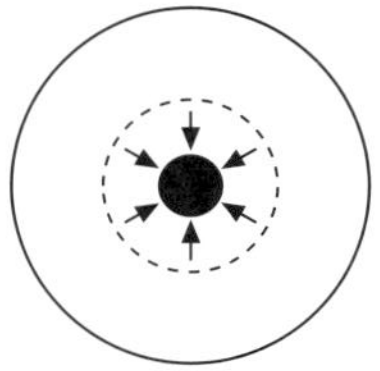

High light intensity: circular muscles of iris contract and the pupil is reduced in size - **less light may enter** and **retina is protected from bleaching**.

Aqueous humour: watery fluid which supports the cornea and the front chamber of the eye.

Vitreous humour: a jelly-like substance which helps to keep the shape of the eyeball, supports the lens and keeps the retina in place at the rear of the eye.

The brain is an integrator - it is able to accept sensory information from a number of sources, compare it with previous experience (learning) and make sure that the appropriate actions are initiated.

Meninges: membranes which line the skull and cover the brain. They help to protect and nourish the brain tissues, but may be infected either by a virus or a bacterium to cause the potentially fatal condition **meningitis**.

Cerebral cortex: has motor areas to control voluntary movement, sensory areas which interpret sensations and association areas to link the activity of motor and sensory regions. The centre of intelligence, memory, language and consciousness.

Forebrain: here many emotions are localised, and damage to this area may cause aggression, apathy, extreme sexual behaviour and other emotional disturbances.

Hypothalamus: contains centres which control thirst, hunger and thermoregulation.

Pituitary gland: is a link between the central nervous system and the endocrine system. Secretes a number of hormones, including follicle stimulating hormone which regulates development of female gametes, and anti-diuretic hormone which controls water retention by the kidney.

Medulla: the link between the spinal cord and the brain, and relays information between these two structure. Has a number of reflex centres which control

a vital reflexes which regulate heartbeat, breathing and blood vessel diameter

b non-vital reflexes which co-ordinate swallowing, salivation, coughing and sneezing.

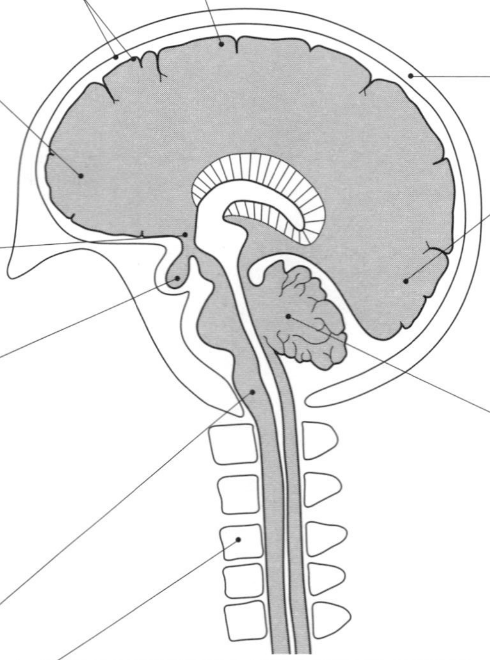

Skull: the cranium is a bony 'box' which encloses and protects the brain.

Visual centre: this area of the cerebral cortex

a interprets impulses along the optic nerve i.e. responsible for vision

b has the connector neurones for both accommodation and the pupil reflex.

Cerebellum: co-ordinates movement using sensory information from position receptors in various parts of the body. Helps to maintain posture using sensory information from the inner ear. Can control learned sequences of activity involved in dancing, athletic pursuits and in the playing of musical instruments.

Spine (vertebral column): composed of 33 separate vertebra which surround and protect the spinal cord; between each pair of vertebrae two **spinal nerves** carry sensory information into the spinal cord and motor information out of it. Dislocation of the vertebrae may compress the spinal nerves, causing great pain, or even crush the spinal cord, leading to paralysis.

DON'T FORGET

- in addition to its being a centre of learning the brain is the relay centre for **cranial reflexes** such as blinking.
- **conditioned reflexes** involve setting up nerve pathways in the brain which link a substitute stimulus to a certain response e.g. hearing a lunch bell may set off the production of saliva.
- the evolutionary advantages of mammals may be partly due to a constant body temperature allowing the brain to 'work' for longer periods so that learning is more successful.

DRUGS AND BRAIN FUNCTION

Alcohol has a progressive effect on the brain

- 1–2 units release inhibitions by affecting emotional centres in the forebrain
- 5–6 units affect co-ordinated movements by influencing motor areas of the cerebral cortex
- 7–8 units cause stupor and insensitivity to pain as sensory areas, including the visual centre, are impaired
- more than 10 units may be fatal as the vital centres of the medulla and hypothalamus are severely inhibited.

Heroin can cause euphoria and insensitivity to pain as pain and emotion centres are inhibited.

L.S.D. may cause hallucinations by altering the balance of brain neurotransmitters and causing loss of the medulla's ability to 'filter' information reaching the cerebral cortex.

Blood sugar regulation is another example of **homeostasis**.

WHAT IS BLOOD SUGAR LEVEL?

Glucose is the cells' main source of energy, and it must always be available to them…

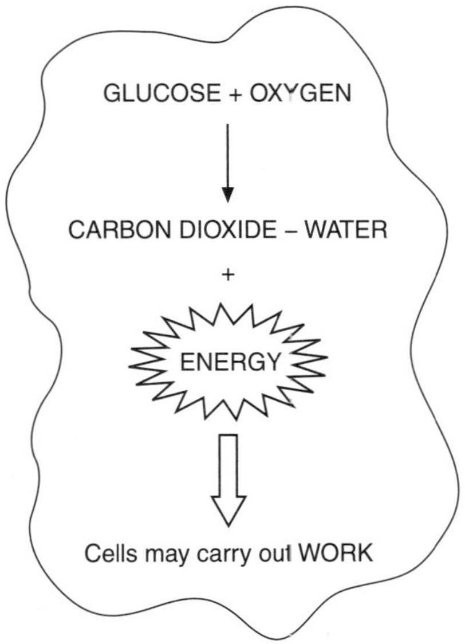

…so that the body keeps a constant amount of glucose in the blood. This is the **blood sugar level** and is usually maintained at about 1 mg of glucose per cm^3 of blood.

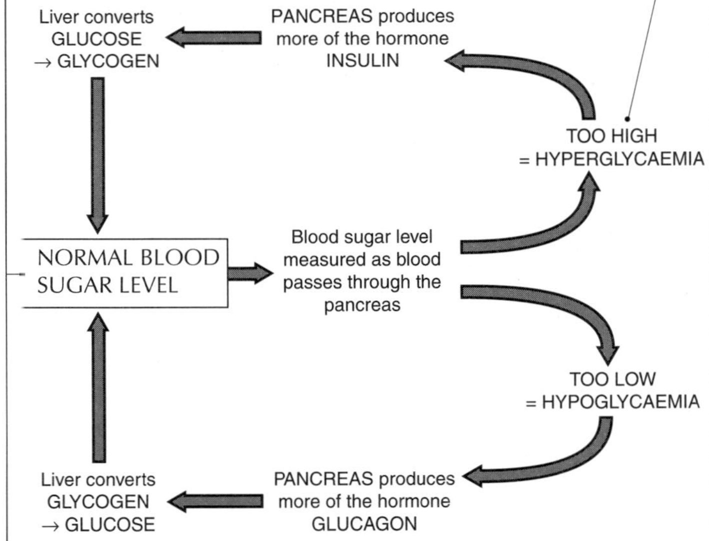

WHAT IS DIABETES?

Diabetes is a condition in which there are higher than normal blood glucose concentrations

- usually the result of the failure of the pancreas to secrete enough insulin
- symptoms include
  excessive thirst, hunger or urine production
  sweet smelling breath
  high 'overflow' of glucose into urine (test with **clinistix**)
- long-term effects if untreated include
  premature ageing
  cataract formation
  hardening of arteries
  heart disease
- treatment is by regular injection of pure insulin - much of this is now manufactured by **genetic engineering**.
- mild diabetes can be controlled by adjusting diet to reduce sugar intake.

Glucose tolerance test measures the body's response to high glucose intake (sugar solution).

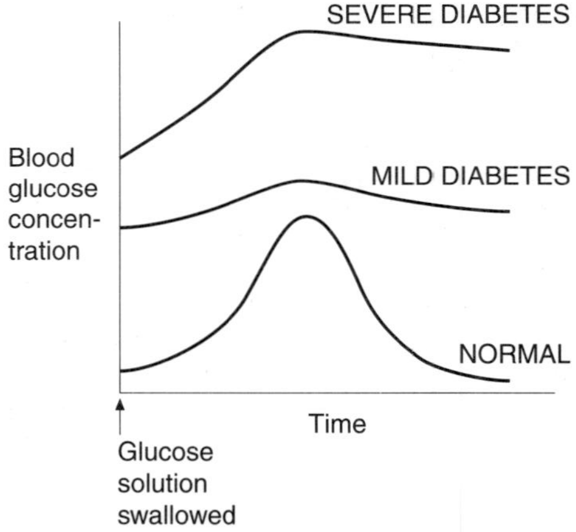

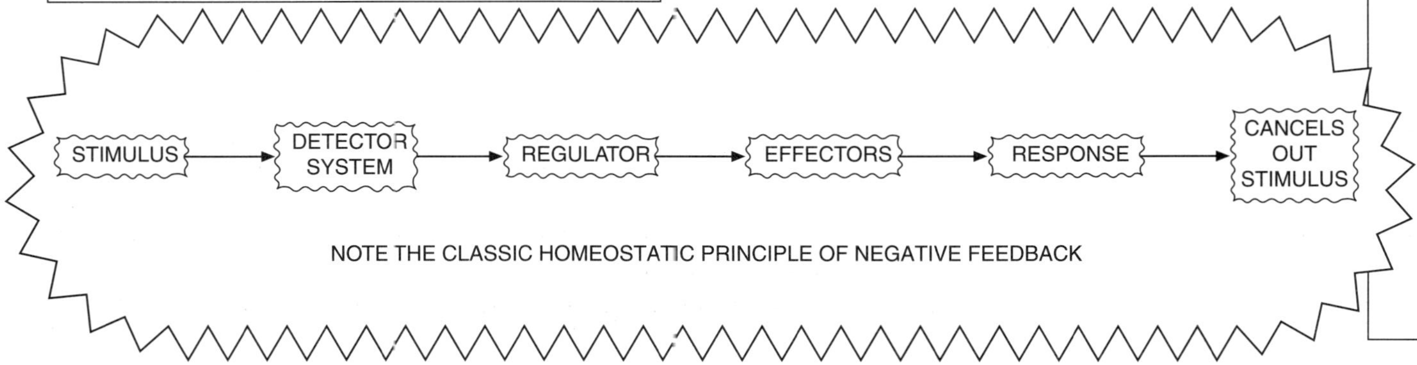

Excretion is removal from the body of the waste products of metabolism.

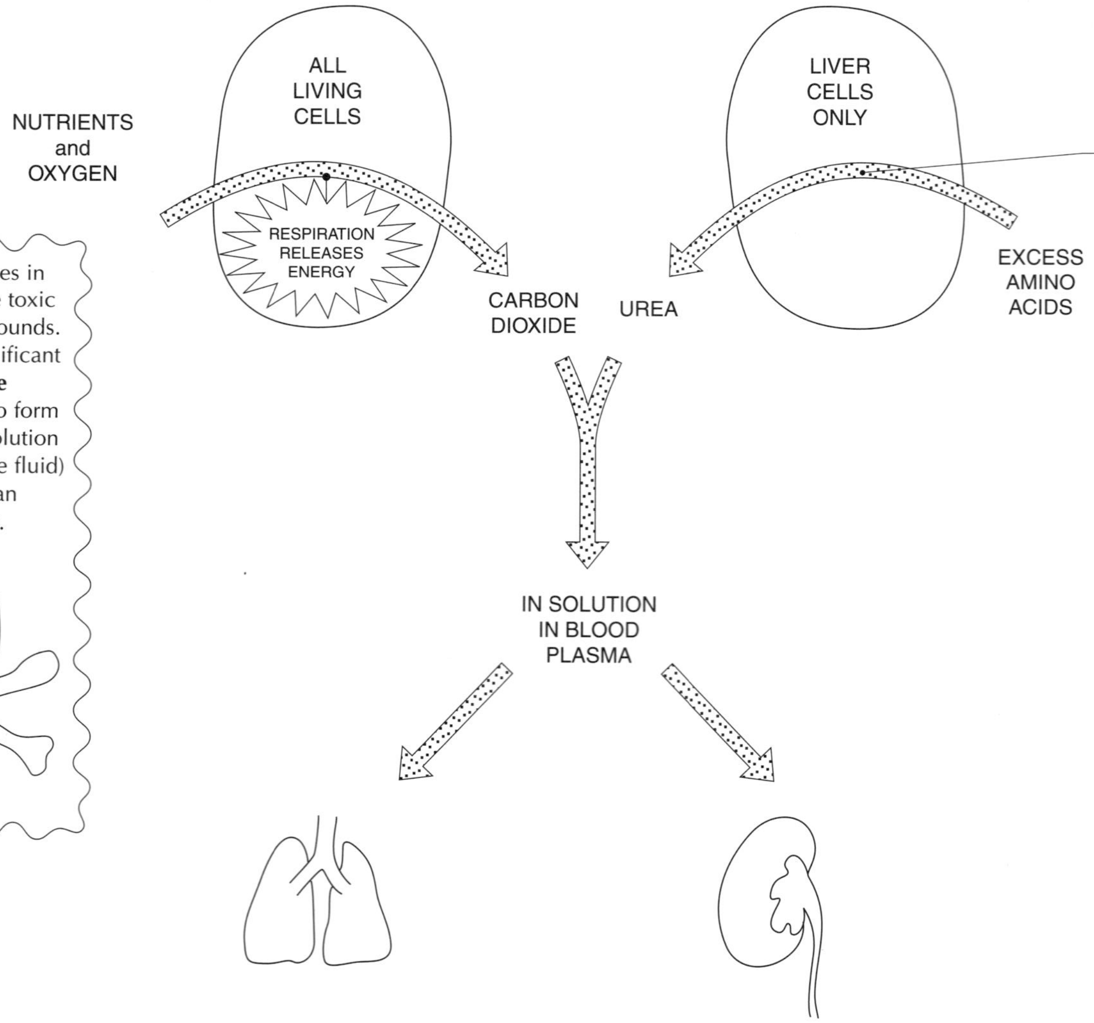

Carbon dioxide is removed from the body by diffusion across the lung surface.

Urea is removed by ultrafiltration of the blood in the kidney, and expelled from the body in the urine.

Metabolic processes in cells may produce toxic (poisonous) compounds. The two most significant are **carbon dioxide** (which dissolves to form a weakly acidic solution in blood and tissue fluid) and **urea** which can denature enzymes.

Deamination is the removal of the amino groups (-NH_2) from excess amino acids, since the body has no storage facility for these compounds.

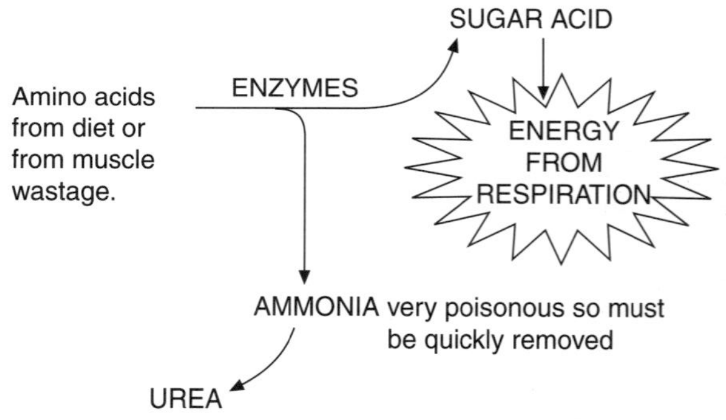

DON'T FORGET!

- **Egestion** is the passage from the body of indigestible substances **which have never been absorbed from the gut**, e.g. dietary fibre.
- Removal of waste products from the body involves **loss of water**. This is a major demand on the body's water.
- Urea is **produced in the liver** but **excreted from the kidney**.
- Metabolic processes also produce **heat**. In excess this is dangerous (causes enzyme denaturation) and is largely removed through the lungs and the skin.

Kidney structure and function

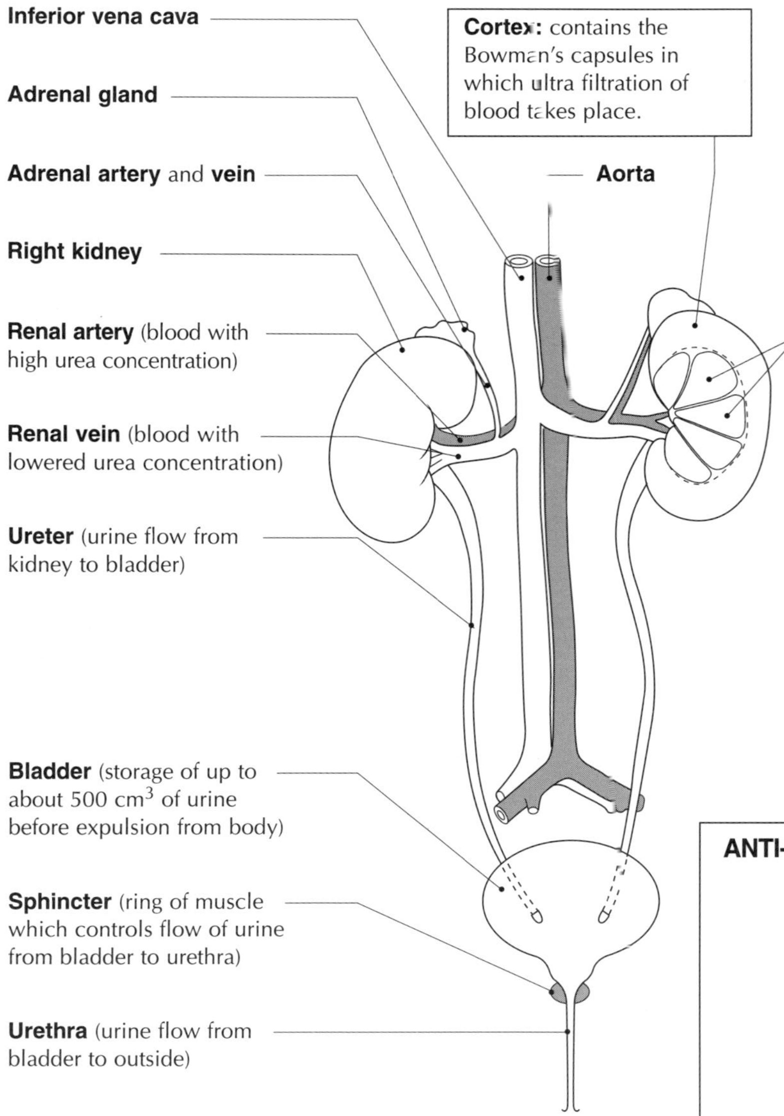

THE KIDNEY IS COMPOSED OF MANY NEPHRONS (KIDNEY TUBULES)

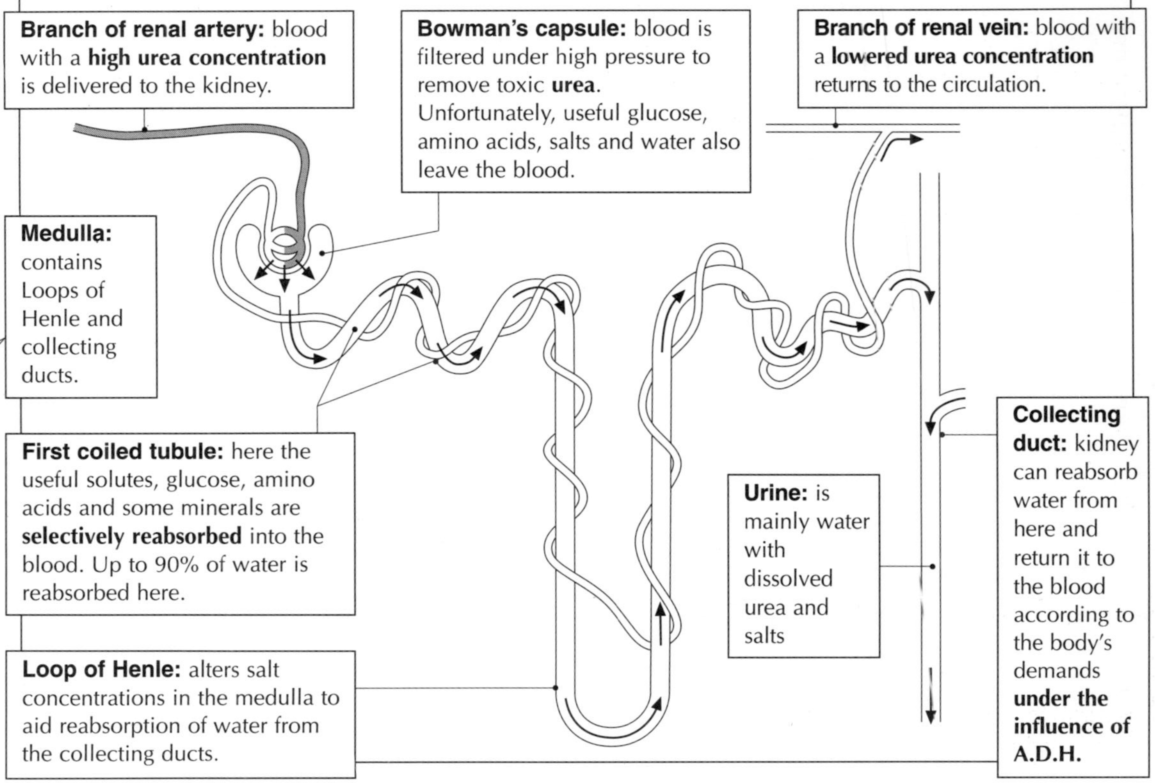

ANTI-DIURETIC HORMONE (A.D.H.) HELPS TO REGULATE BODY WATER

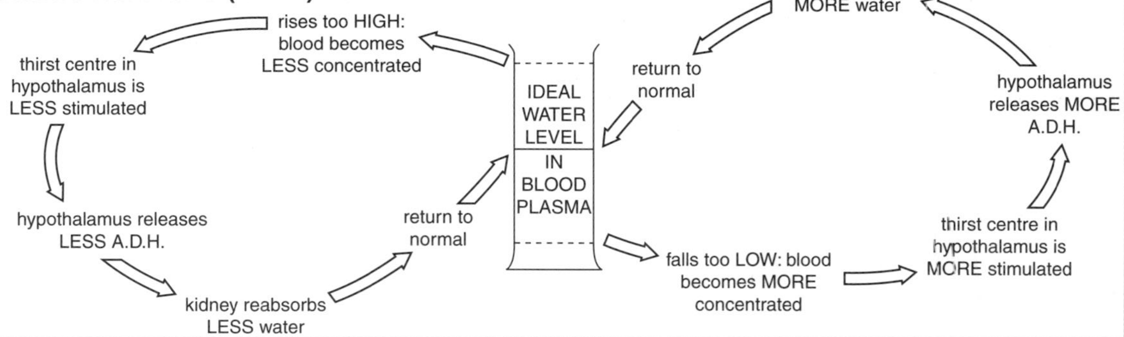

Kidney failure: if one or both kidneys fail then dialysis is used or a transplant performed to keep urea and solute concentration in the blood constant.

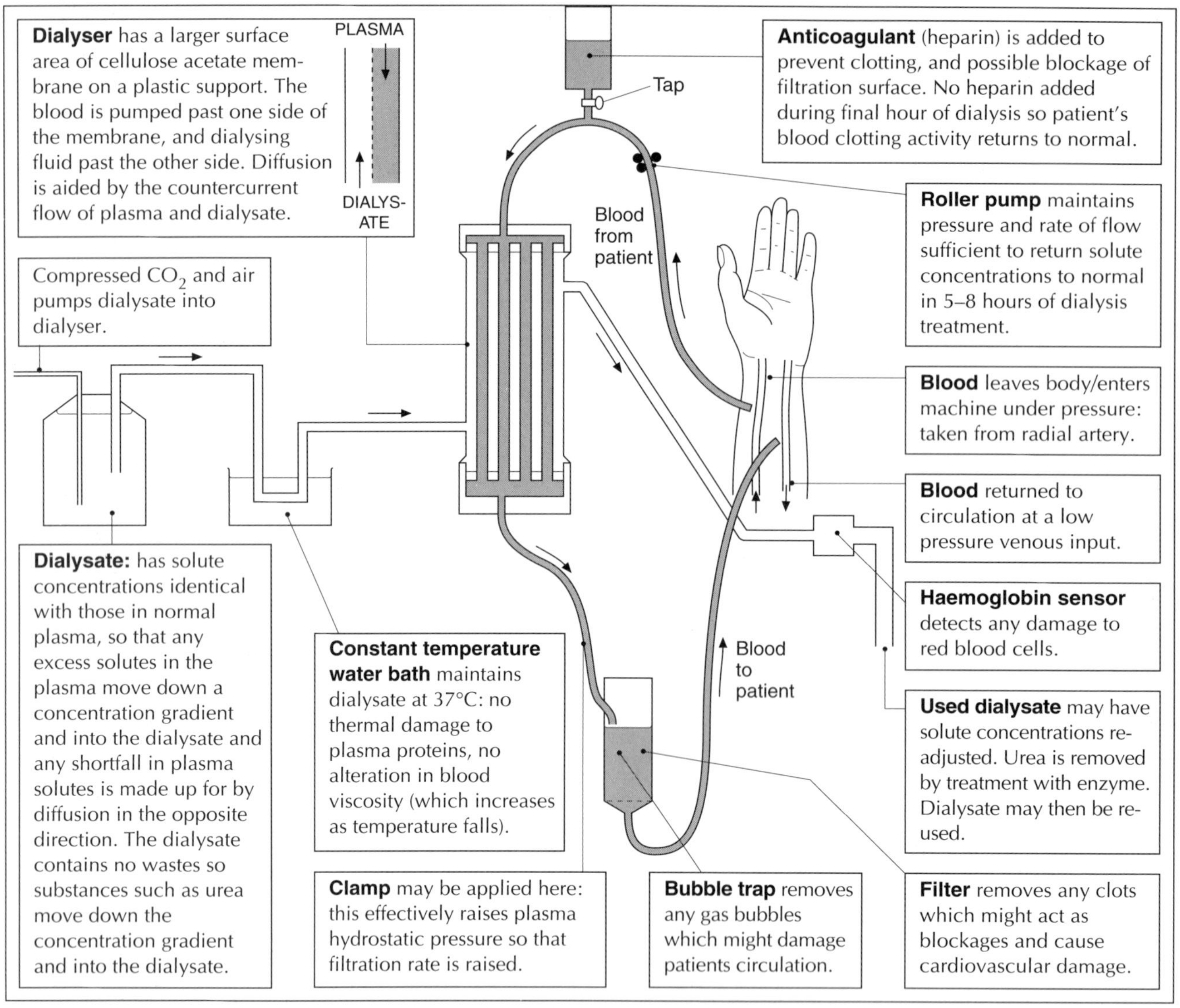

KIDNEY TRANSPLANTATION

may be necessary as renal dialysis is inconvenient for the patient and costly.

Kidney transplants have a high success rate because:

1 the vascular connections are simple

2 live donors may be used, so very close blood group matching is possible

3 because of 2 there are fewer immuno-suppression-related problems in which the body's immune system reacts against the new kidney.

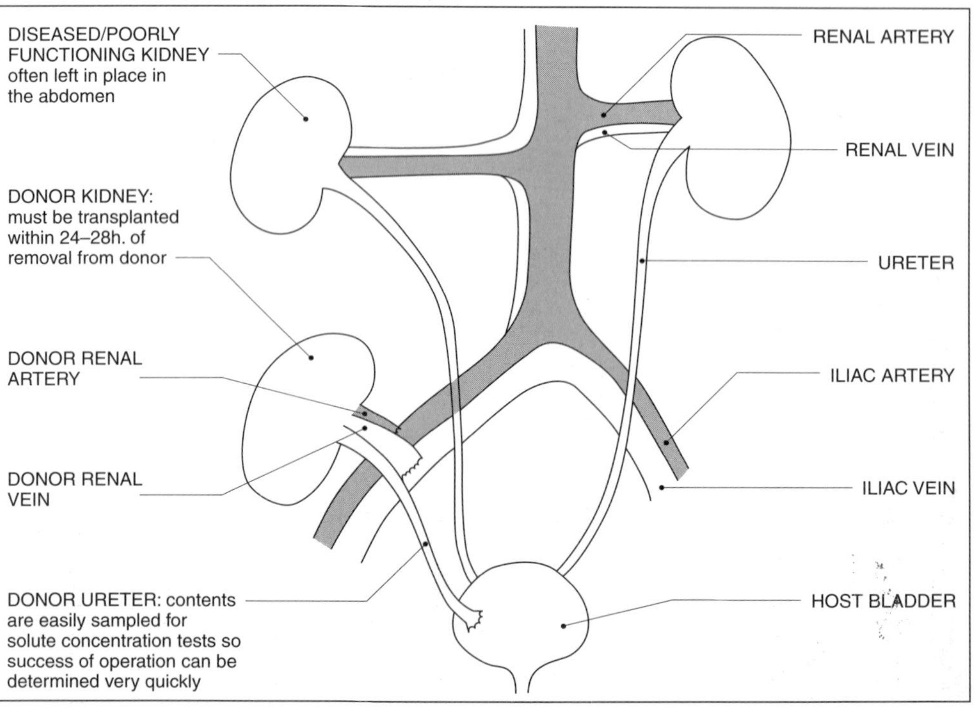

Temperature control in endothermic organisms

is an example of **negative feedback**. Any deviation from optimum conditions sets mechanisms in motion to cancel out the change.

The **temperature control centre** is located in the **hypothalamus** in the floor of the brain. It receives incoming messages about the body temperature and makes certain that the appropriate corrective mechanisms operate to cancel out any deviations.

Skin increases heat loss

- Sweat is secreted and **evaporation** consumes heat from body.
- Vasodilation of surface capillaries allows blood to **radiate** heat away from body.
- Relaxation of hair erector muscles allows hair to lie flat against skin to permit maximum heat loss by convection.

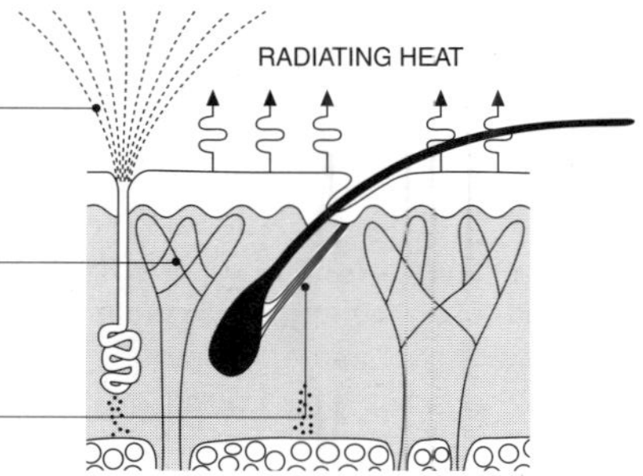

The **optimum temperature** for the body's activities is about 37°C. This ensures that

- enzymes work rapidly but are not denatured
- cell membranes keep their structure
- blood does not become too viscous to pass through narrow capillaries.

NORM → INCREASE → DETECTOR → INTEGRATOR → CORRECTIVE MECHANISMS: attempt to increase heat loss → NORM

NORM → DECREASE → DETECTOR → INTEGRATOR → CORRECTIVE MECHANISMS: attempt to conserve heat → NORM

Deviations from the norm are detected

a. by **hot** and **cold sensors** in the skin

b by sensors in the hypothalamus which measure blood temperature.

During **fever**, the body's temperature may be reset 2-3°C higher: the higher temperatures denature proteins in the infecting bacteria more rapidly than they affect human proteins.

Hypothermia ("low temperature")

- occurs if body's core temperature falls below 35°C
- causes irrational behaviour, brain damage, circulatory problems
- elderly people are very susceptible because they may be immobile
- children are very susceptible because they have a high surface area to volume ratio.

Humans may also

shiver - muscular activity generates heat

add extra clothing - insulation reduces heat loss

eat more - eating stimulates heat production by respiration…

…or **turn on the central heating!**

(humans can control their **external** environment too!)

- Contraction of hair erector muscles traps a layer of still air which reduces heat losses by **convection**.
- Vasoconstriction of surface capillaries shunts blood away from skin and so reduces heat losses by **radiation**.
- Sweat glands do not secrete sweat so there is no consumption of body heat in **evaporation**.

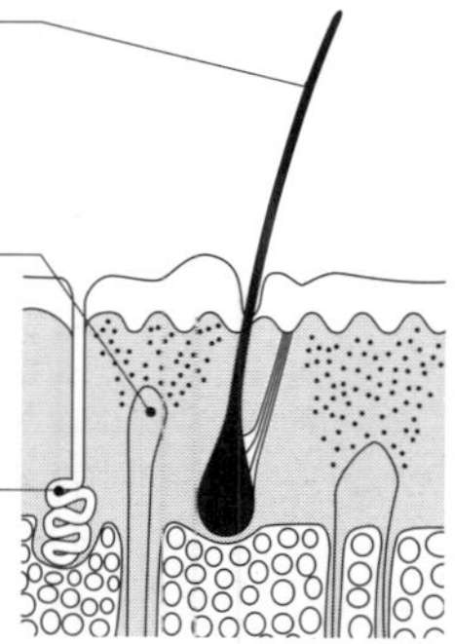

SKIN REDUCES HEAT LOSS

Muscle-bone machines

are responsible for movement.

Antagonistic pairs of muscles are necessary for controlled movement at a joint. Muscles may only exert a force by **contraction**. To reverse a muscular movement therefore requires contraction of an opposing (antagonistic) muscle.

Human forelimb

Bending the arm:
Biceps contracts
Triceps relaxes

Straightening the arm:
Triceps contracts
Biceps relaxes

So, the biceps is a **flexor** and the triceps is the **extensor** of the elbow joint, and biceps and triceps make up an **antagonistic pair**.

Tendons connect muscle to bone. They are **inelastic** so that muscle contraction can be converted into movement of bone.

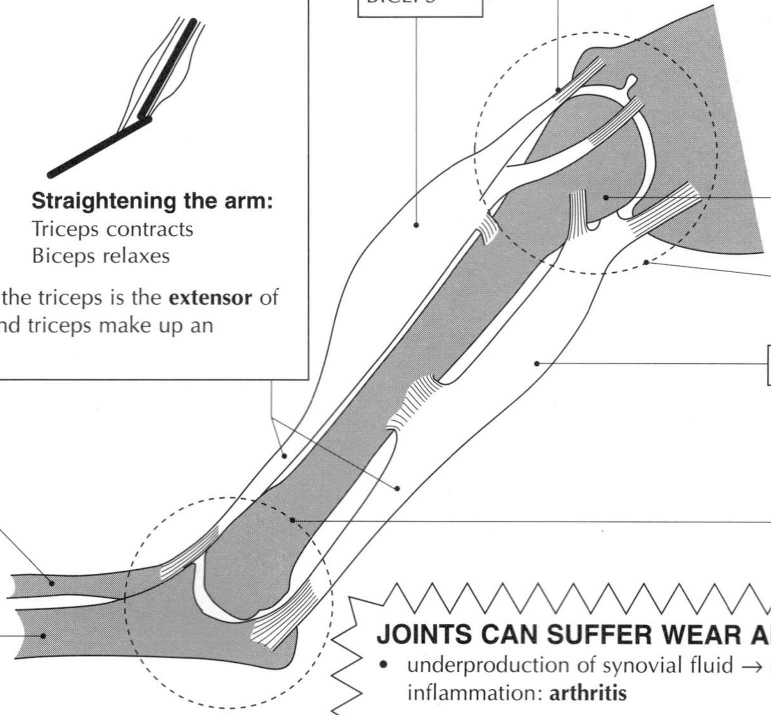

CONTRACTION OF MUSCLE REQUIRES

- the **supply** of **glucose** and **oxygen** to release energy, in the form of ATP, by respiration
- the **removal** of **carbon dioxide** and **heat** (if respiration becomes **anaerobic** there must also be the removal of **lactic acid** or cramp may result)

These requirements are satisfied by the development of a system of capillary beds in the muscle together with their arteries and veins.

- a **stimulus** in the form of a **nervous impulse** delivered from the end plates of **motor neurones**.

Synovial joints are freely-moving because of the arrangement of the synovial membranes/cartilage between the ends of adjacent bones.

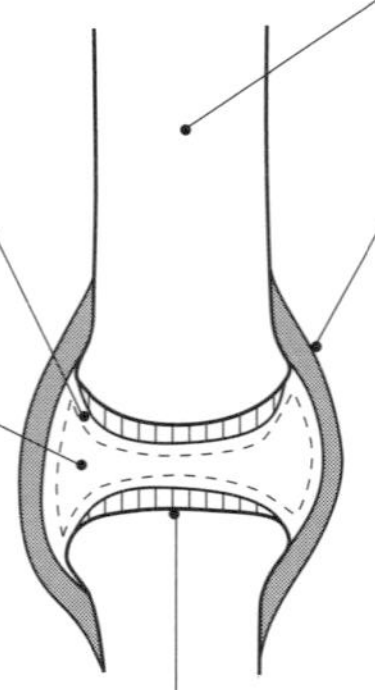

Bone

Synovial membrane secretes the synovial fluid

Synovial fluid looks and feels like egg-white. It lubricates the joint and contains phagocytic cells which remove debris resulting from wear and tear in the joint.

Ligaments are elastic but strong. They **allow movement** at the joint but help to **prevent dislocation**. Ligaments connect bone to bone.

Cartilage covers the ends of the two bones which meet at a joint. It **reduces friction** between the bones and helps to **absorb shock** if the joint is compressed.

Ball-and-socket joints (e.g. hip and shoulder) allow free movement in many planes.

Hinge joints (e.g. elbow, knuckles) allow free movement in one plane only.

JOINTS CAN SUFFER WEAR AND DAMAGE

- underproduction of synovial fluid → stiffness and inflammation: **arthritis**
- dislocation may occur if ligaments are torn of overstretched
- cartilage may be displaced following impact on a joint. The cartilage fragments may impede movement and may need to be removed.
- **replacement joints** may be used if original is damaged/severely arthritic. These are made from stainless steel (strong, non-toxic, non-corrosive) and may be coated with 'teflon' on the working surface.

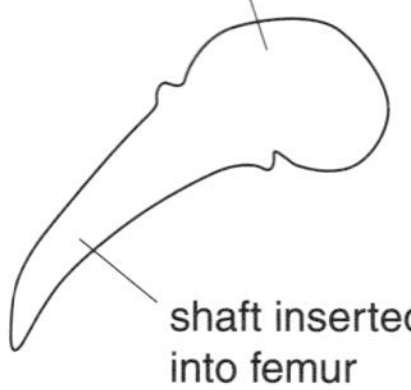

OTHER FUNCTIONS OF THE BONY SKELETON

- **support** - air itself is not dense enough to support the body of a land mammal
- **protection** - soft tissues may be enclosed inside a bony case e.g. brain inside the skull.
- **synthesis** - red and white blood cells are produced in bone marrow inside long bones such as the femur.

MALE REPRODUCTIVE SYSTEM

- produces **testosterone**
- manufactures **male gametes**
- transfers male gametes to female

Development of **male secondary sexual characteristics** under control of the hormone **testosterone**.

Development of **sex organs** started by **pituitary hormones** and kept going by **testosterone**.

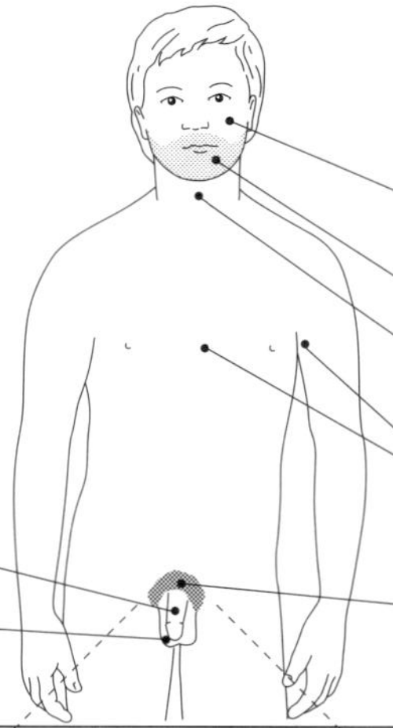

Skin becomes more oily - may cause ACNE

Facial hair develops

Changes to larynx - DEEPER VOICE

Hair grows on chest and in armpits

Pubic hair develops

Penis grows

Testes grow - produce sperm and testosterone

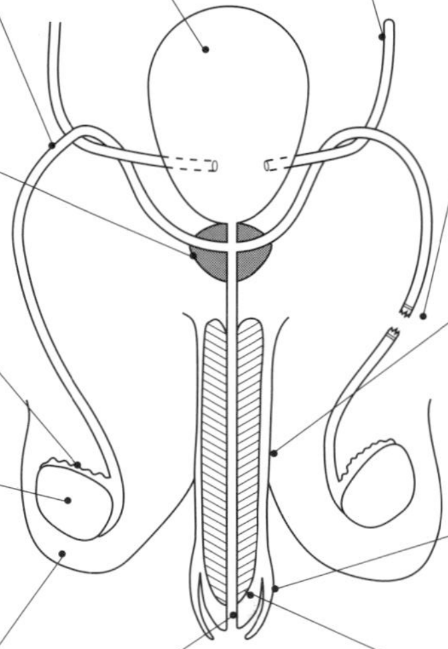

Sperm duct (vas deferens) pathway for sperm transport (propelled by peristalsis) to tip of penis.

Prostate gland produces part of fluid component of semen.

Epididymis is the site of sperm storage.

Testis is the site of sperm manufacture and production of testosterone.

Scrotum is a sac of skin hanging outside the abdomen and holding testes at a temperature (2 or 3°C below body temperature) which is ideal for sperm production.

Urethra is a tube for both semen and urine (during intercourse a valve closes the exit from the bladder).

Foreskin covers and protects the sensitive tip of the penis.

Penis with erectile tissue can be stiffened by filling with blood. Used to deliver sperm to the vagina of the female (**ejaculation**).

Male sterilisation (vasectomy): the sperm duct is cut and tied so no sperm can be ejaculated. It does not inhibit hormone production by the testes so should not diminish sex drive.

Condom (sheath): a thin rubber covering fitted over the erect penis. Collects sperm and prevents them reaching the vagina. Also offers protection against sexually transmitted diseases such as gonorrhoea and AIDS.

CONTRACEPTION PREVENTS PREGNANCY

FEMALE REPRODUCTIVE SYSTEM

- produces **oestrogen/progesterone**
- develops **female gametes**
- accepts sperm
- allows development/birth of foetus

Development of **female secondary sexual characteristics** under the control of the hormone **oestrogen**.

Development and function of **sex organs** largely controlled by **pituitary hormones**.

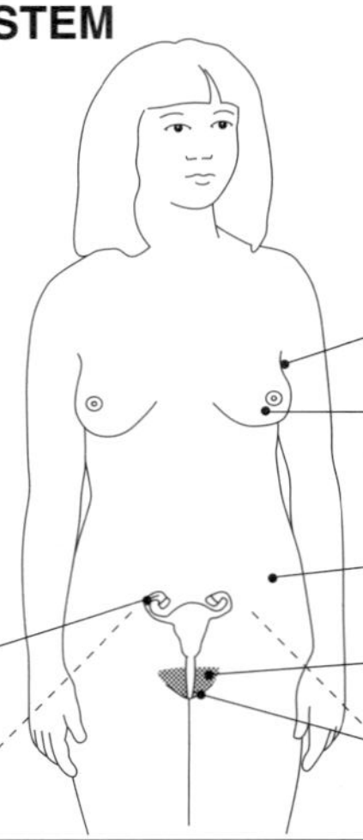

Hair grows in armpits

Breasts develop/increase in size

Hips broaden

Pubic hair grows

Vagina enlargens

Ovaries grow - eggs develop and female hormones are released

Female sterilisation: the oviducts are cut or tied so that no eggs can be passed down to be fertilised.

Coil (I.U.D. - Intra-Uterine Device): plastic with copper wire which irritates the lining of the uterus and prevents implantation of a fertilised egg.

Cap (diaphragm): a thin rubber shield fitted over the cervix before intercourse. It prevents entry of sperm.

Spermicide (cream): is placed inside the vagina before intercourse - kills sperm. Usually used in conjunction with the diaphragm.

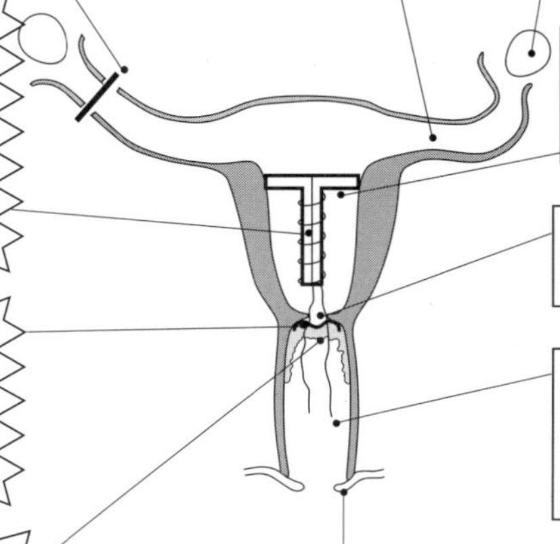

Oviduct: eggs travel down this tube to the uterus - fertilisation usually occurs in the first third of the oviduct.

Ovary: site of egg development and production of oestrogen and progesterone.

Uterus: site of implantation of embryo and development of foetus.

Cervix: sperm deposited here during intercourse.

Vagina: accepts the penis during intercourse. Baby passes down the vagina during birth.

Vulva: thick folds of skin which surround and protect the opening of the vagina.

The Menstrual cycle

The Menstrual cycle is a regular series of changes to the female reproductive system in preparation for fertilisation and pregnancy. It is controlled by **hormones** from the **pituitary gland** and the **ovary**.

In humans the cycles of the two ovaries are out of phase, so **each ovary** ovulates every 56 days but **each female** ovulates (releases an egg) every 28 days.

Separation of uterus lining - blood and fragments of tissue leave the body via the vagina. This monthly 'period' occurs only in primates such as humans.

The uterus lining begins to degenerate **unless embryo implantation** has occurred when **progesterone** (from the corpus luteum) keeps the lining intact to begin pregnancy.

Uterus lining and its blood vessels are well-developed to receive an embryo. This optimum set of conditions remains for 6-7 days after ovulation.

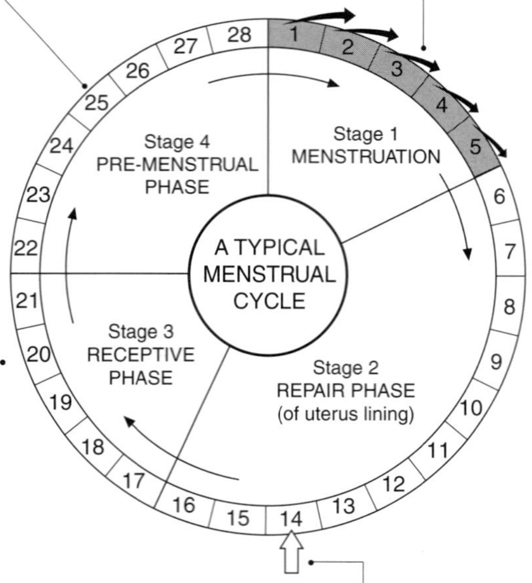

Ovulation, the release of the egg from the Graafian follicle into the oviduct. This is stimulated by **luteinising hormone** and the release of fluid and an associated 'blip' in body temperature (see right) means that some human females are aware that ovulation has occurred.

Contraception and the menstrual cycle

- **The pill:** an oral dose of one or both of the hormones **oestrogen** and **progesterone** which acts as a **feedback inhibitor** of the release of luteinising hormone by the pituitary gland. Thus **ovulation cannot occur** and **no pregnancy can result**.
- **The rhythm method** assumes that the time up to two days before ovulation and from four days after ovulation are **safe**, i.e. there should be no eggs available to be fertilised. This method is unreliable.

THE MENSTRUAL CYCLE IS CONTROLLED BY HORMONES

Brain - receives information from other parts of the body, processes it and then 'instructs' pituitary gland.

Pituitary gland - releases hormones which control activity of the ovary.

Follicle stimulating hormone - stimulates development of Graafian follicle in the ovary.

Luteinising hormone - stimulates release of mature ovum from ovary.

Luteinising hormone - stimulates development of corpus luteum from the remains of the follicle.

Oestrogen - repairs the lining of the uterus and stimulates development of female sexual characteristics.

Progesterone - keeps the lining of the uterus ready for implantation and pregnancy.

The lining of the uterus gradually thickens and develops more blood vessels ready to receive a fertilised egg.

This 'blip' in temperature corresponds to ovulation.

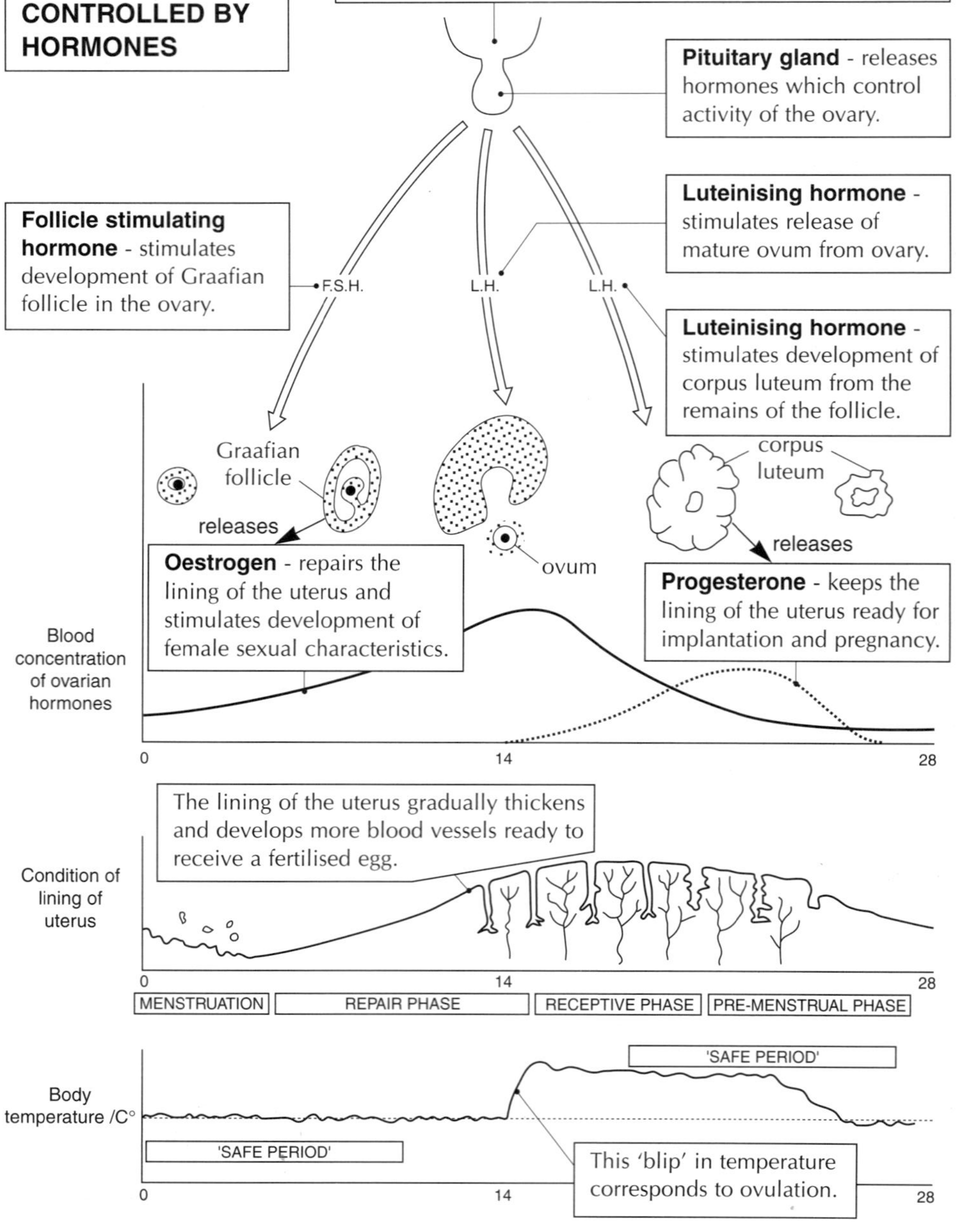

The placenta is responsible for protection and nourishment of the developing foetus.

PLACENTA: this disc-shaped organ has a number of functions:

- exchange of soluble materials between mother and foetus
- physical attachment of the foetus to the wall of the uterus
- protection (1) of foetus from mother's immune system
  (2) against dangerous fluctuations in mother's blood pressure
- hormone secretion. These hormones keep the wall of the uterus in 'pregnancy' state as the corpus luteum breaks down by the 3rd month.

The placenta is lost as the 'afterbirth' following birth of the foetus.

UMBILICAL CORD: contains blood vessels which carry materials which will be / have been exchanged between mother and foetus. The cord connects the foetus to the placenta.

Wall of uterus: this is very muscular. At full term of the pregnancy a hormone (OXYTOCIN) is secreted from the pituitary gland which makes this muscle contract in a series of waves to expel the foetus. The same hormone is used in the intravenous drip which induces birth when the pregnancy has gone on for too long.

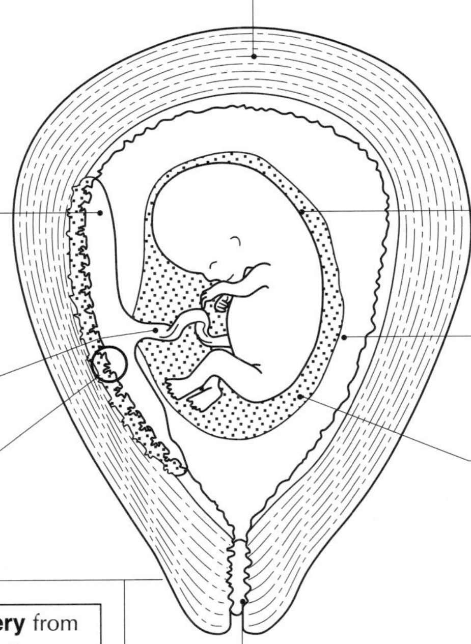

Foetus: this develops from a **zygote**.

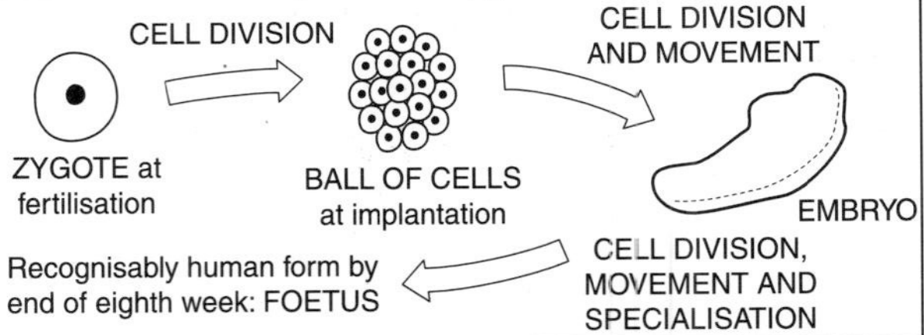

Amnion: the membrane which encloses the amniotic fluid. This is 'ruptured' just before birth.

Amniotic fluid: protects the foetus against

- mechanical shock
- drying out
- temperature fluctuations

Some of the foetal cells fall off into this fluid and can be collected by **amniocentesis**. The cells can be analysed to detect disease, genetic abnormalities and even the sex of the foetus.

Mucus plug in cervix: protects foetus against possible infection. The plug is expelled just before birth.

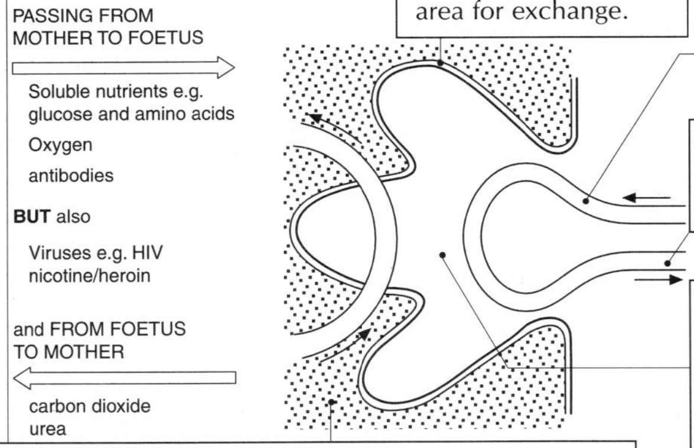

Placental villi - increase the surface area for exchange.

Umbilical artery from foetus to placenta - deoxygenated blood containing waste.

Umbilical vein from placenta to foetus - oxygenated blood cleared of waste.

Counter-current flow system blood in maternal and foetal blood vessels **flows in opposite directions**. This gives the maximum possible area over which concentration gradients favour diffusion.

'Pit' in wall of uterus contains mother's blood. N.B. there is **no direct contact** between maternal and foetal blood.

GROWTH AND DEVELOPMENT OF THE FOETUS

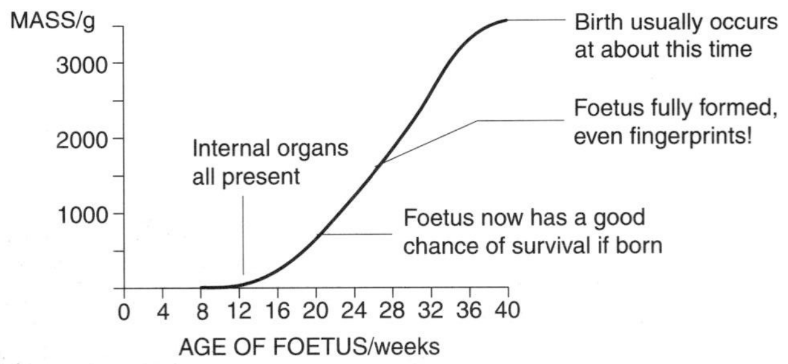

- Complete period from fertilisation to birth = **gestation period**.
- Cell division converts single cell (zygote) to 30 million million in a newborn baby.
- Most rapid growth from the twelfth week. As much as 1500x gain in mass in 20 weeks.

Plants make food by photosynthesis

LEAVES ARE ADAPTED FOR PHOTOSYNTHESIS

Leaves are **thin**: few layers of cells so light can reach the sites of photosynthesis.

Large **surface area** to trap the maximum amount of sunlight.

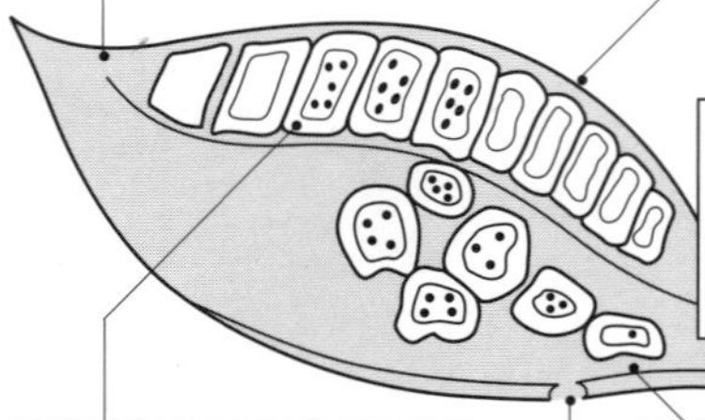

Vascular bundles transport water and mineral salts in (**xylem**) and sugars out (**phloem**).

Palisade layer near to upper surface has many chloroplasts in cells tightly packed for maximum light absorption.

Spongy layer has loosely-packed cells, covered with a thin film of water to allow rapid diffusion of carbon dioxide.

Stomata are pores which allow the entry of **carbon dioxide** and the exit of **oxygen**. The pores can be opened and closed by the action of guard cells. This conserves water since water easily evaporates from the surface of spongy cells.

The occurrence of photosynthesis can be demonstrated by testing for the presence of starch (made from glucose).

STARCH —(IODINE SOLUTION)→ INTENSE BLUE-BLACK COLOURATION

The rate of photosynthesis can be measured by collecting evolved oxygen, usually from the cut end of the aquatic plant, *Elodea*.

EQUATION FOR PHOTOSYNTHESIS

CARBON DIOXIDE + WATER ⟹ GLUCOSE + OXYGEN (LIGHT ENERGY; CHLOROPHYLL, ENZYMES)

$$6\,CO_2 + 6H_2O \longrightarrow C_6H_{12}O_6 + 6O_2$$

LIMITING FACTORS AND PHOTOSYNTHESIS

There are a number of conditions which must be satisfied for photosynthesis to proceed. The factor which **is nearest to its minimum value** will limit the rate of this process.

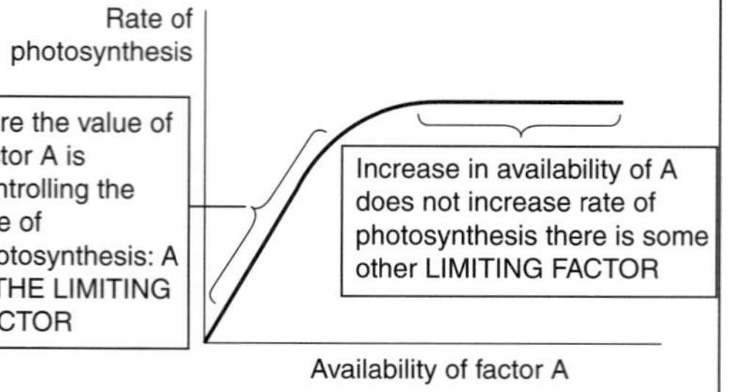

Limiting factors in photosynthesis are

carbon dioxide concentration

light intensity

temperature - affects the enzymes which catalyse the chemical reactions of photosynthesis.

APHIDS TAP PHLOEM FOR THE SUGARS MADE BY PHOTOSYNTHESIS

WHAT HAPPENS TO THE GLUCOSE?

Glucose

- Respiration to provide the energy to drive the metabolic reactions needed to keep the plant alive.
- Conversion to other molecules such as oils and proteins. This may require mineral salts (e.g. nitrate).
- Conversion to sucrose for transport to other parts of the plant via the phloem.
- Conversion to cellulose for the construction of plant cell walls.
- Conversion to starch for storage. Starch is insoluble so does not affect water potential of plant cells.

CHEMISTRY OF PHOTOSYNTHESIS: photosynthesis is a **two stage process**. Both stages occur in chloroplasts.

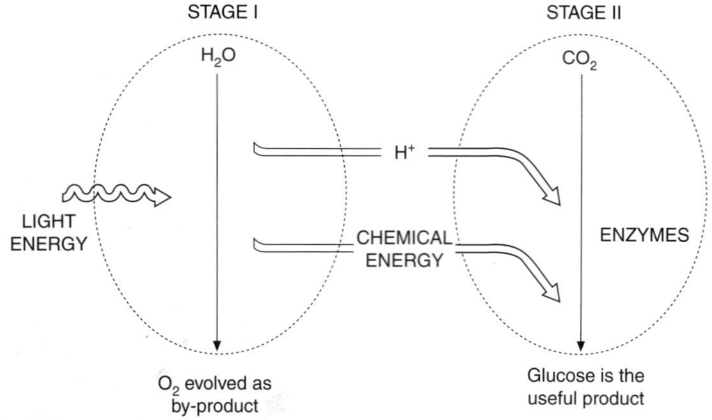

Transport systems in plants

move vital substances from **sources** (sites of uptake or manufacture) to **sinks** (sites of production or storage).

STEM: the position of the vascular bundles (in a ring with soft cortex in the centre) helps to support the stem against sideways forces, e.g. wind.

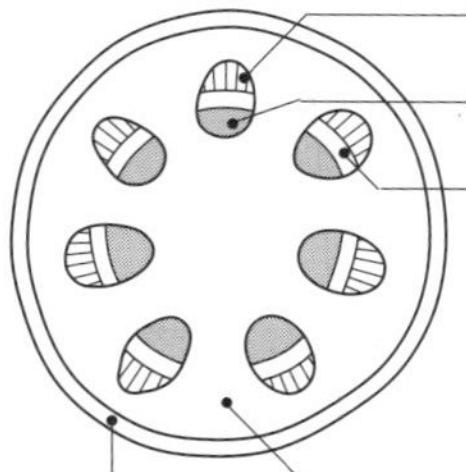

- **Phloem**
- **Xylem**
- **Cambium:** cells which divide to provide more phloem and xylem

} vascular bundle

Cortex: cells become turgid (filled with water) and help to support non-woody parts.

Epidermis: protective against, for example, infection by viruses and bacteria, and dehydration.

ROOT:

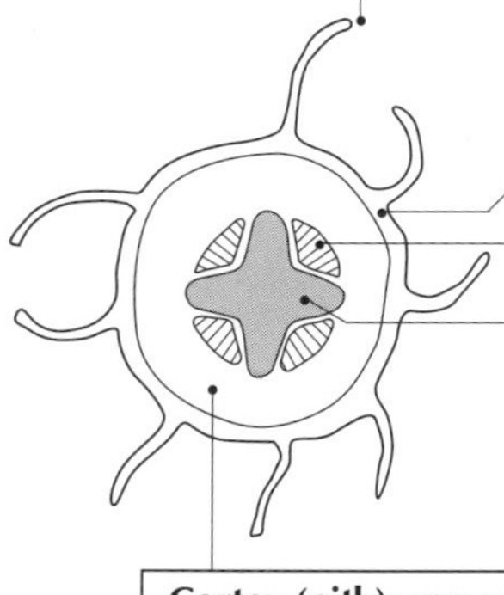

Root hair: extended cells of epidermis increase surface area for water/ion uptake.

Epidermis: protective, e.g. against infection by fungi.

Phloem: } **Xylem:** } together form a strong central rod which helps to prevent root being pulled out of the soil.

Cortex (pith): can act as a winter store for starch.

INVESTIGATION OF FUNCTIONS OF PLANT TISSUES

- **Phloem:** aphids (greenfly) can sample contents and radioactive sugars identify tissue as site of sugar transport.
- **Xylem:** water-soluble dyes trace pathway of water movement
- **Roots:** inhibitors of respiration stop active uptake of ions

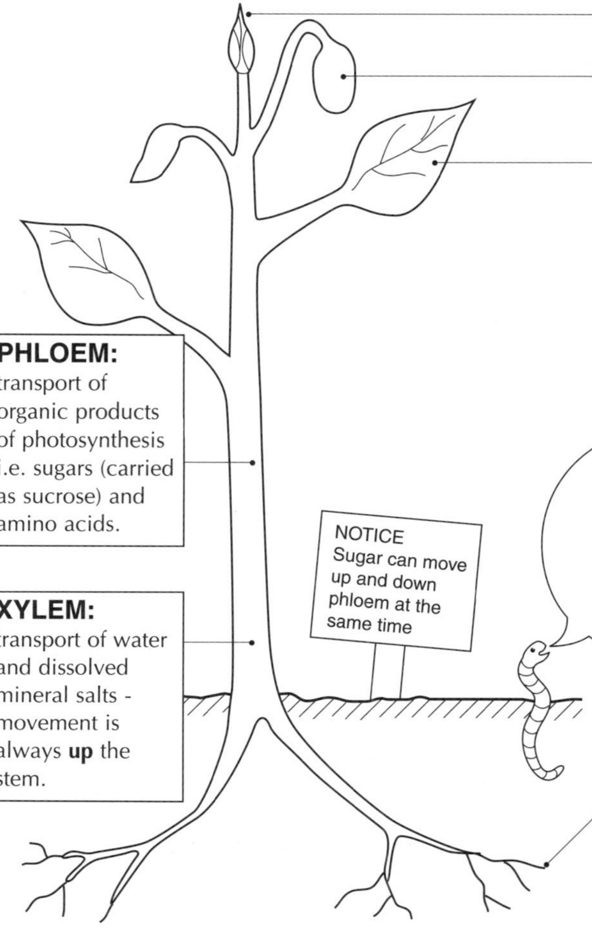

PHLOEM: transport of organic products of photosynthesis i.e. sugars (carried as sucrose) and amino acids.

XYLEM: transport of water and dissolved mineral salts - movement is always **up** the stem.

FRUITS AND GROWING POINTS: these are **sinks** for many nutrients.

Fruits

- demand water for swelling of ovary wall if succulent
- demand sucrose to be converted to starch as an energy store

Growing points

- demand water for cell expansion
- demand sucrose as energy source for cell division
- demand all nutrients as raw materials for cell production.

LEAVES: are both **sinks** and **sources**.

Sinks

- water as a reactant in photosynthesis
- magnesium as a component of the chlorophyll molecule
- sucrose, when leaves are young, as a source of energy and subunits of cellulose.

Sources

- glucose formed during photosynthesis

  CARBON DIOXIDE + WATER
  ↓
  GLUCOSE + OXYGEN
- useful ions, e.g. magnesium, just before leaf fall.

DIRECTION OF TRANSPORT VARIES WITH THE SEASONS!
Sucrose will be transported **from** stores in the root **to** leaves in Spring, but **to** stores in the root **from** photosynthesising leaves in the Summer and early Autumn.

ROOTS: are both **sinks** and **sources**.

Sinks

- sucrose to supply energy for growth and active uptake of ions from the soil
- sucrose to be converted to starch for storage

Sources

- water, absorbed from the soil solution
- ions, absorbed from soil by active transport
- sucrose, before leaves are capable of photosynthesis, in Spring.

Water movement through a plant

begins with evaporation from the leaf surface.

WATER LOSS FROM LEAVES can be demonstrated using cobalt chloride paper.

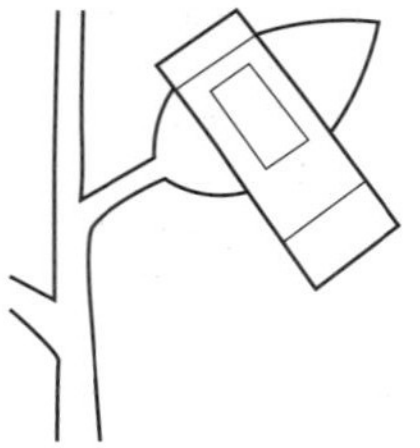

Anhydrous cobalt chloride paper (**blue**) turns **pink** as water evaporates from leaf.

WATER UPTAKE BY PLANTS can be measured using a bubble potometer.

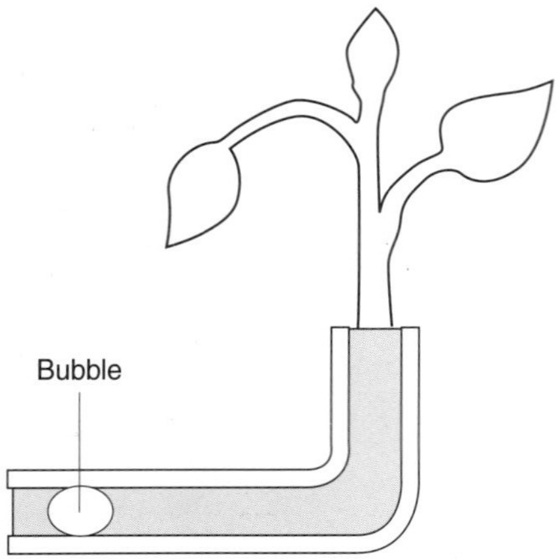

Water loss draws bubble along capillary tube as water uptake replaces loss.

WATER IS IMPORTANT
- it is a substrate for photosynthesis
- it is a pathway for transport of mineral ions through the plant
- it provides support, through turgidity
- evaporation may help to cool leaf surfaces

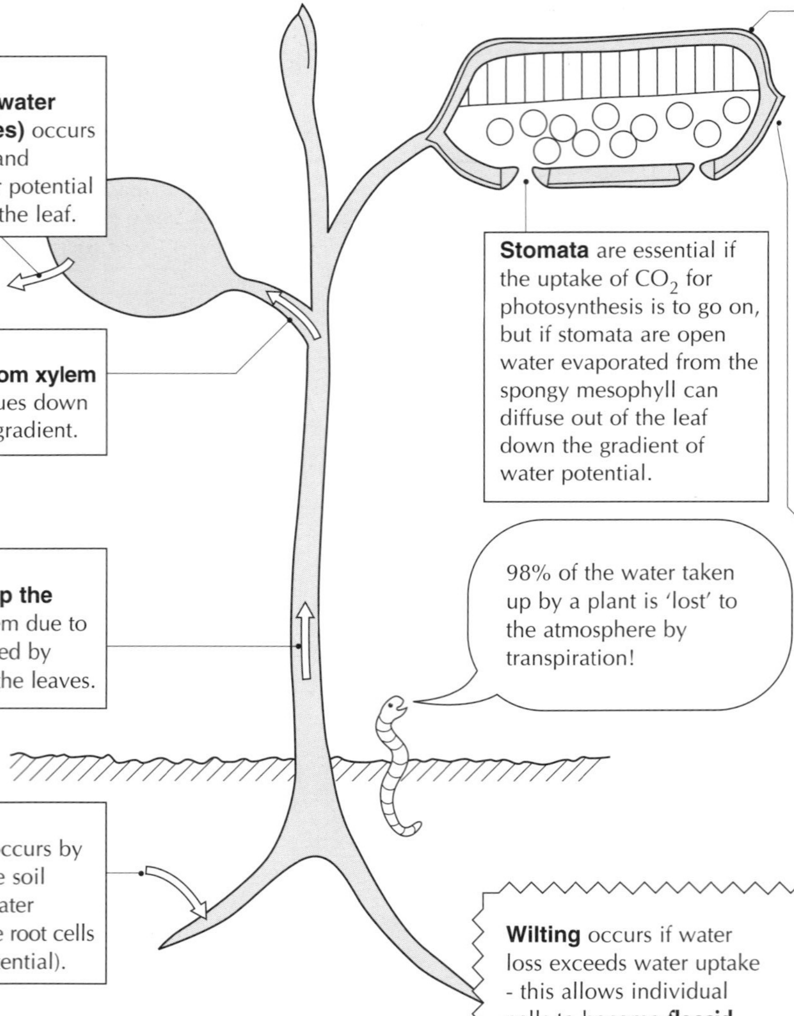

1
Transpiration (water loss from leaves) occurs by evaporation and lowers the water potential in the tissues of the leaf.

2
Water moves from xylem to enter leaf tissues down water potential gradient.

3
Water moves up the stem in the xylem due to the **tension** caused by water loss from the leaves.

4
Water uptake occurs by osmosis from the soil solution (high water potential) into the root cells (lower water potential).

Flooding of soil with salt water lowers the water potential of the soil solution and prevents osmosis - no water uptake by roots.

Stomata are essential if the uptake of CO_2 for photosynthesis is to go on, but if stomata are open water evaporated from the spongy mesophyll can diffuse out of the leaf down the gradient of water potential.

98% of the water taken up by a plant is 'lost' to the atmosphere by transpiration!

LEAF STRUCTURE MAY REDUCE TRANSPIRATION
- thick, waxy cuticle reduces evaporation from epidermis.
- stomata may be sunk into pits which trap a pocket of humid air.
- leaves may be rolled with the stomata on the inner surface close to a trapped layer of humid air.

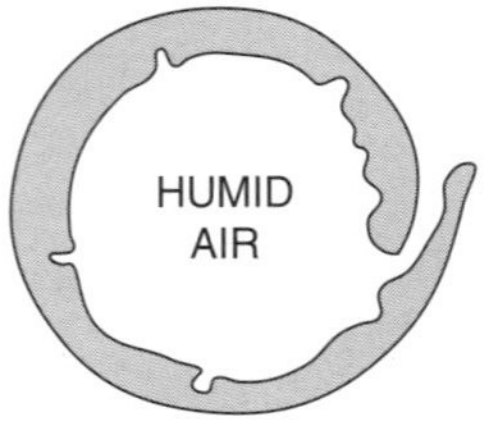

- leaves may be needle-shaped to reduce their surface area.

ATMOSPHERIC CONDITIONS MAY AFFECT TRANSPIRATION
- **Wind** moves humid air away from the leaf surface and increases transpiration.
- **High temperatures** increase the water-holding capacity of the air and increase transpiration.
- **Low humidity** increases the water potential gradient between leaf and atmosphere and increases transpiration.
- **High light intensity** causes stomata to open (to allow photosynthesis) which allows transpiration to occur.

Wilting occurs if water loss exceeds water uptake - this allows individual cells to become **flaccid**, the tissues become limp and the plant is no longer supported.

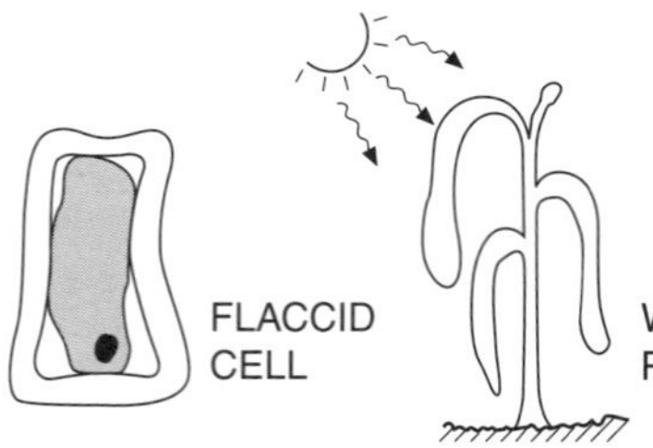

Hormones and minerals affect plant growth and development

HORMONES

SHOOTS ARE POSITIVELY PHOTOTROPIC - they show **growth movements towards light**.

- **Mechanism:** the plant growth substance called **auxin** is inhibited by light. It therefore accumulates on the shaded side of shoots **causing faster growth on the shaded side** and thus **'bending' towards light**.

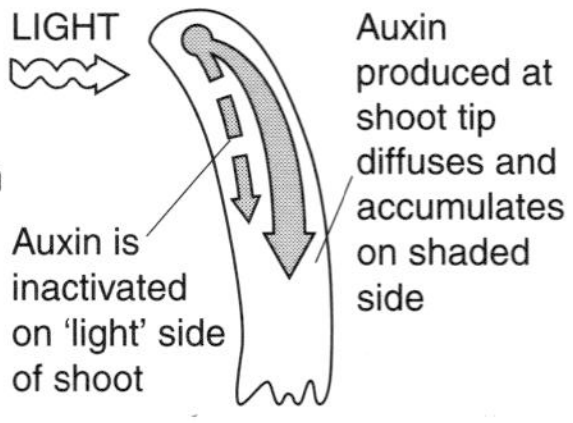

- **Benefit:** leafy shoots tilt towards light to improve rate of photosythesis and therefore increase rate of growth.

PLANT HORMONES HAVE COMMERCIAL USES

Plant hormones can be sprayed onto flowers to mimic fertilisation and increase the chance of co-ordinated development of fruits.

At the correct concentration plant hormones can stimulate root formation in cuttings.

Plant hormones can act as selective weedkillers since they disrupt the growth pattern of weed species at a concentration which does not affect cereal crops.

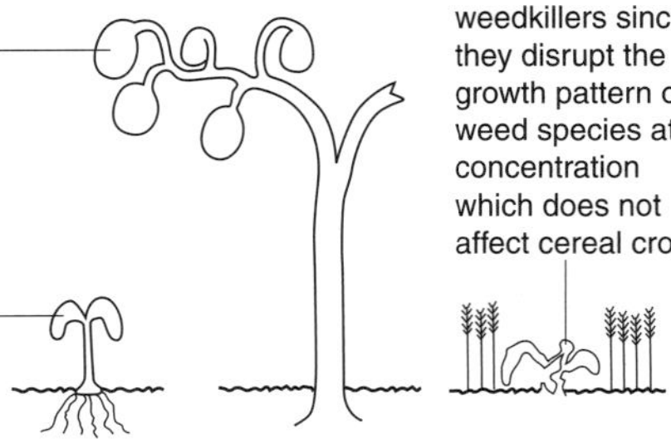

ROOTS ARE POSITIVELY GEOTROPIC - they show **growth movements towards gravity**.

- **Mechanism:** gravity causes **auxin** to accumulate on lower side of root. Unlike shoot growth, root growth is **inhibited** by this concentration of auxin so that **lower side grows more slowly** and root **'bends' towards gravity**.
- **Benefit:** root grows into soil to anchor plant firmly and improve uptake of minerals and water.

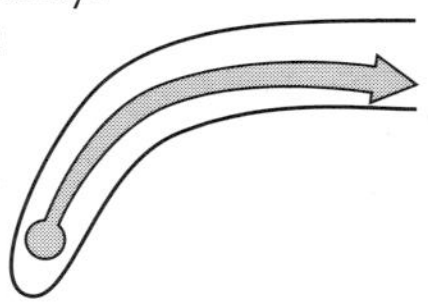

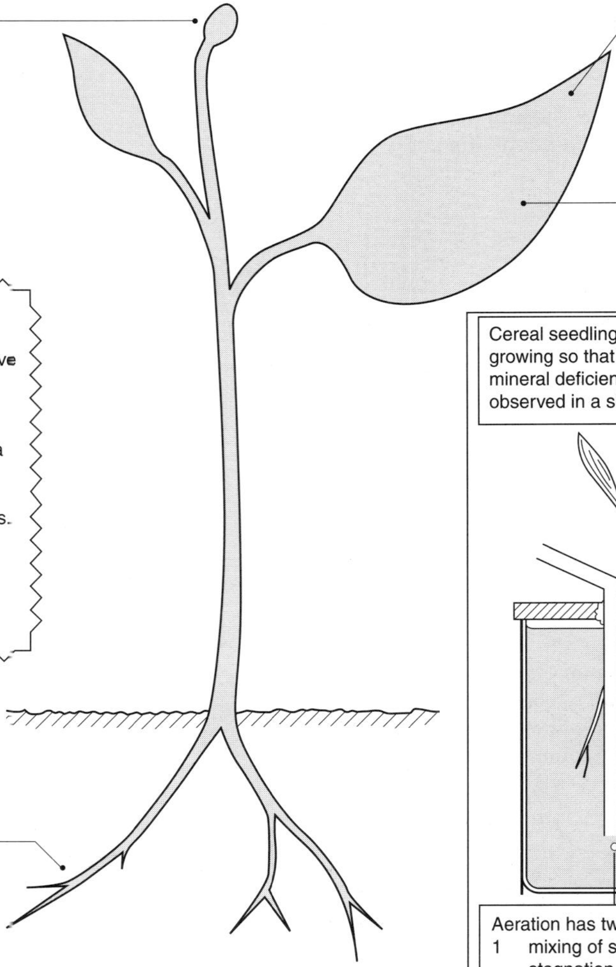

MINERALS

MAGNESIUM is absorbed from the soil solution as the ion, Mg^{2+}. It is required for **manufacture of chlorophyll** - absence causes leaves to turn pale yellow and eventually halts photosynthesis.

NITROGEN is absorbed from soil solution as nitrate (NO_3^-) or ammonium (NH_4^+) ions. It is required for **manufacture of proteins**, **nucleic acids** and **plant hormones**. Absence causes poor growth of all organs, especially leaves. Availability of nitrogen is probably the **major limiting factor** in plant growth in cultivated soils so many farmers add **nitrogen-containing fertilisers** to their land.

WATER CULTURE EXPERIMENTS allow study of plant mineral requirements in controlled conditions which eliminate the many variables associated with soil as a growth medium.

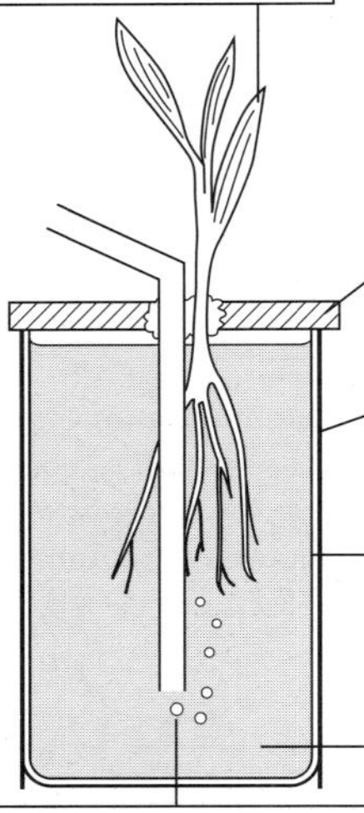

Cereal seedling is rapid-growing so that effects of mineral deficiency are observed in a short time.

Light-proof cover (1) prevents entry of airborne contaminants and (2) supports seedling in growing position.

Black card or foil cover prevents entry of light so that no aquatic organisms can compete with seedling roots for mineral ions. Cover can be easily removed to examine the growth of the seedling roots.

Glass container can be thoroughly cleaned (using acid) so that no minute traces of minerals remain where they might lead to erroneous results.

Growth/nutrient solution contains mineral ions in previously determined optimum concentrations. Complete (control) solution contains all ions. Test solutions have a single ion omitted. NB to eliminate one ion, a salt is omitted but its 'second' ion is replaced.
e.g. for nitrogen-free solution omit calcium nitrate but increase concentration of another calcium salt.
The complete solution is often called Knop's solution.

Aeration has two functions:
1. mixing of solution so that no stagnation occurs
2. oxygenation so that aerobic respiration may provide energy for active uptake of ions from solution.

Flower structure is adapted to pollination

CARPEL: the female part of the flower.

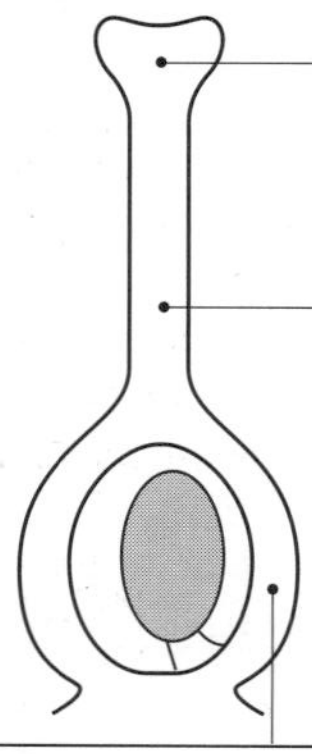

Stigma: surface on which pollen grains, containing male gametes, may be deposited.

Style: stalk which holds stigma in prominent position and down which pollen tube may grow.

Ovary: contains the ovule which encloses the female gamete. Ovary wall may become part of the fruit.

STAMEN: the male part of the flower.

Anther: produces pollen grains, containing male gametes, within the pollen sacs.

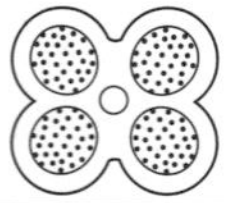

Filament: stalk which holds anther in position to release pollen.

PETALS: play an important part in pollination (see opposite).

SEPALS: protect the other floral parts against drying out and fungal attack. There are the same number as there are petals, and they are usually green in colour.

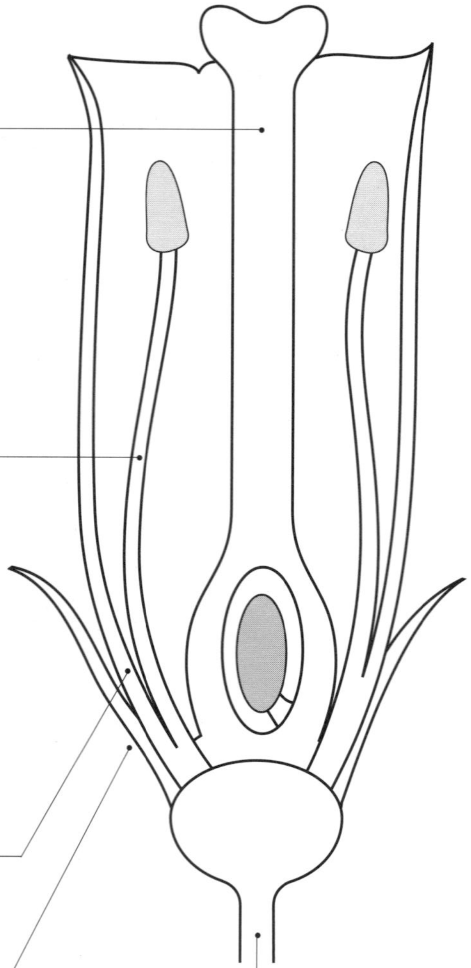

RECEPTACLE: the swollen tip of the flower stalk. It is the base on which the other parts of the flower stand.

POLLINATION is the transfer of pollen from the stamen to the stigma.

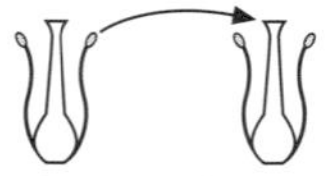

Cross-pollination involves pollen transfer between different flowers of the same species. It involves genetic variation but may be risky because it requires a **vector** (a carrier).

Self-pollination involves pollen transfer from stamen to stigma of the same flower. It is less risky but it limits the chances of genetic variation.

Wind-pollinated species usually occur in dense groups e.g. the grasses.		**Insect-pollinated species** are usually solitary or in small groups.
Dull in colour. Small, or even absent to reduce obstruction of pollen access to stigma.	PETALS	Large, brightly-coloured, may be scented and/or have guidelines. Base may produce attractive **nectar**.
Long and flexible so that pollen may be easily released.	STAMENS	Short and stiff to brush pollen against body of visiting insect.
Long and feathery, giving a large surface area to receive pollen.	STIGMA	Held inside petals to ensure contact with body of visiting insect.
Small, dry, enormous quantities.	POLLEN	Large, sticky, small amounts.

Bees are adapted to feed on pollen and nectar. They pollinate flowers when feeding.

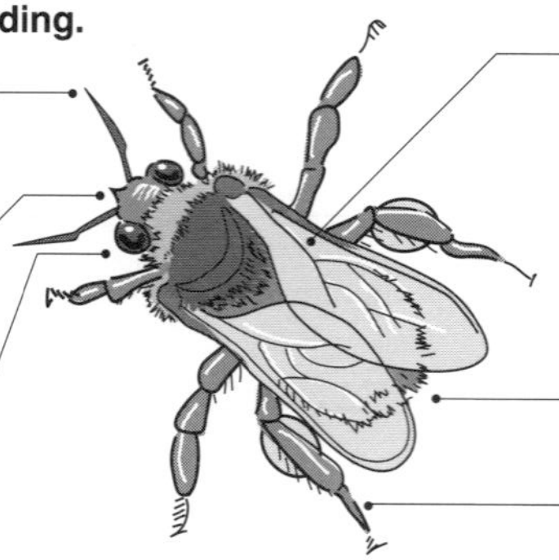

Antennae - detect scent of flower.

Mouthparts can suck nectar from flower.

Eyes - detect colour of flower.

Wings permit movement between flowers.

Hairy abdomen may hold 'sticky' pollen.

Pollen brushes and baskets on third pair of legs collect pollen for larvae in hive.

Fertilisation, fruits and seed dispersal

Pollen grain: chemical signals ensure that it can only germinate (produce a pollen tube) on the stigma of a flower of the same species.

Pollen tube: grows down through the style and acts as a channel to deliver the male gamete from the pollen grain to the female gamete in the ovule.

Ovule: enclosed within the ovary, and contains the female gamete which has the haploid (n) number of chromosomes.

Fertilisation occurs when the male and female gametes, both of which are haploid, fuse to form a diploid (2n) zygote. At the same time other changes occur which convert some of the ovule cells into food reserves.

Male gamete: has the haploid (n) number of chromo-somes and was formed by meiosis (reduction division) in the anther.

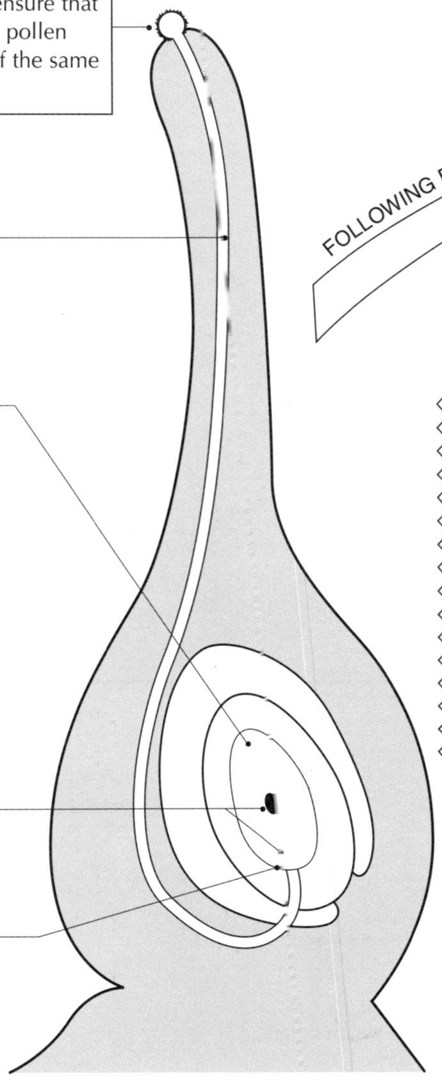

FOLLOWING FERTILISATION

The CARPEL ⟶ FRUIT
with the
OVULE ⟶ SEED

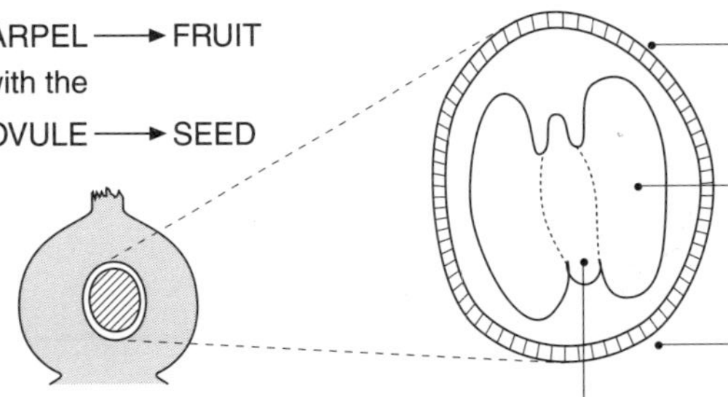

Testa (seed coat): prevents drying out of the embryo.

Cotyledons: form the food store for the embryo - may be **one** in monocotyledons like grasses, or **two** in the dicotyledons like peas.

Micropyle: allows entry of water as seed begins to germinate.

Embryo: following germination this grows and develops into a new young plant.

The fruit may be **adapted to aid dispersal of seeds**

BY WIND

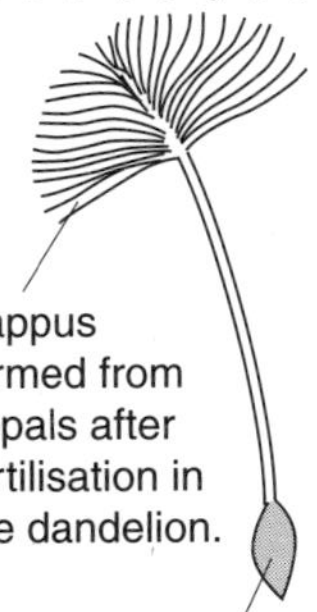

Pappus formed from sepals after fertilisation in the dandelion.

Seed contained in two fused carpels.

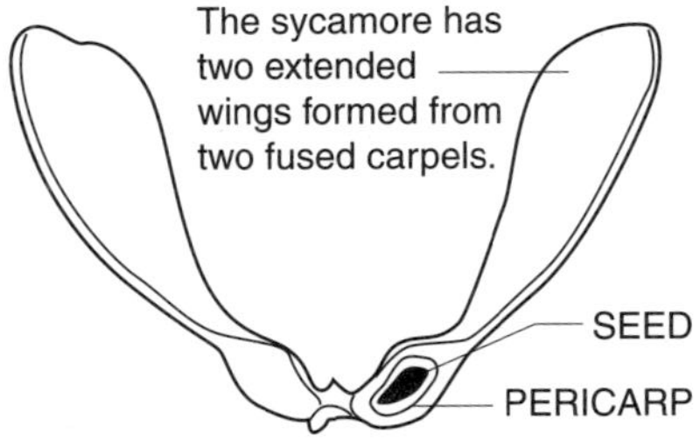

The sycamore has two extended wings formed from two fused carpels.

BY ANIMALS

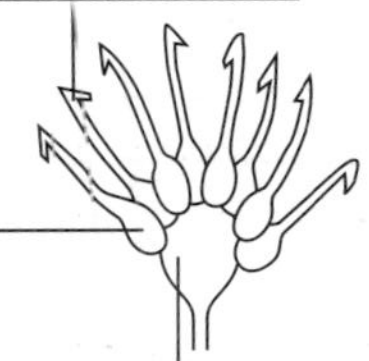

Hooked style is lignified (woody) and attaches to hair or fur of animals.

Ovary contains seed.

Avens is a false fruit since the style (not just the ovary) is involved in dispersal

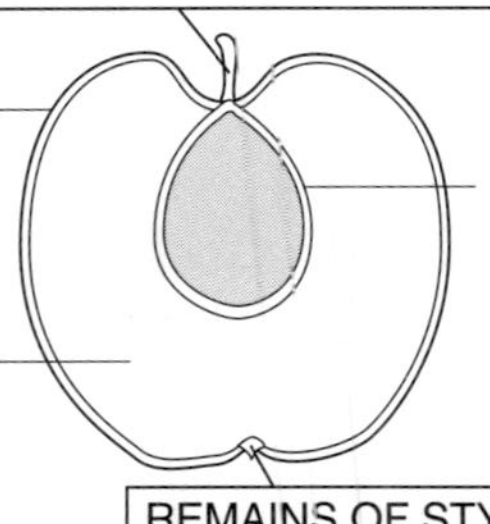

Pedicel - the “stalk” which attached the fruit to the adult plant.

Coloured skin attracts animals.

Succulent flesh of plum offers ‘reward’ to animals.

BY WATER

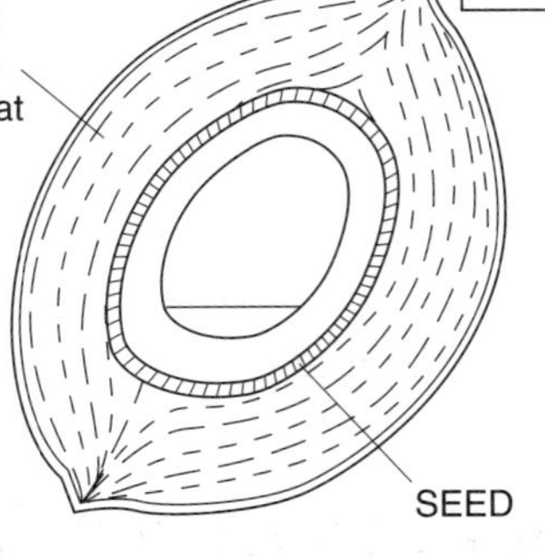

The fibrous ovary wall of the coconut enables the fruit to float in water so that the seed is water-dispersed.

Keys and classification

*** the FIVE KINGDOMS**

A key enables identification of an organism by observation of its characteristics. Close observation allows a series of questions (the branch points in this key) to be answered, eventually leading to the organism being studied.

LIVING ORGANISMS

- (1)
  - made up of single cells
    - (2)
      - cells have no clear nucleus — **BACTERIA** e.g. *Salmonella* *
      - cells have a clear nucleus — **PROTOCTISTA** e.g. *Plasmodium* *
  - made up of many cells
    - (3)
      - cells have no cell wall — **ANIMALS** *
        - (4)
          - organism has hard exoskeleton, jointed limbs, segmented body — ARTHROPODS
            - (5)
              - Three body segments (head, thorax, abdomen), Three pairs of legs, Two pairs of wings, Antennae — INSECTS e.g. Mosquito
              - Two pairs of antennae — CRUSTACEA e.g. Crab
              - Two body segments (head-thorax, abdomen), Four pairs of legs, No wings — SPIDERS e.g. Tarantula
          - (6) segmented body with chaetae (bristles) — ANNELID e.g. Earthworm
          - (7) soft body with no limbs - covered by shell — MOLLUSC e.g. Snail
          - organism has internal skeleton, non-segmented body — VERTEBRATES
            - (8)
              - No covering on skin (smooth, moist skin) — AMPHIBIANS e.g. Frog
              - Skin covering
                - (9)
                  - Scales on skin
                    - (10)
                      - Dry skin — REPTILE e.g. Lizard
                      - Moist (mucus-covered skin), Gills, Fins — FISH e.g. Stickleback
                  - No scales on skin
                    - (11)
                      - Feathers, Beak, Wings — BIRD e.g. Thrush
                      - Fur/hair, Mammary glands — MAMMAL e.g. Human
      - cells have obvious cell wall
        - (12)
          - cells contain chlorophyll in chloroplasts (so organism feeds by photosynthesis) — **PLANTS** *
            - (13)
              - no separate root, stem and leaves — ALGAE
              - stem, leaves but no roots — MOSSES
              - stem, leaves and roots
                - (14)
                  - Spores produced — FERNS
                  - Seeds produced — ANGIOSPERMS e.g. Oak tree
          - cells do not contain chlorophyll (so organism feeds by absorption) — **FUNGI** e.g. Bread mould *

What are you? Follow the branch points at 1, 3, 4, 8, 9, and 11 to identify yourself as a **mammal**.

Naming and classifying living organisms

THE BINOMIAL SYSTEM OF NOMENCLATURE

- provided by Linnaeus
- successful since
  (a) each species has a unique name
  (b) shows which species are closely related e.g. *Panthera leo* and *Panthera tigris*
- incorporates both species and genus into specific name
  e.g. *Homo sapiens*
  (i) Underline or *italicise*
  (ii) Capital letter for genus, lower case for species.
- the 'naming' of organisms (usually after they have been placed into 'groups' or taxons)
- needed to aid communication between groups of scientists so 'name' must be unambiguous and easily understood
- usually in Latin since
  (a) original scientific language
  (b) universally accepted

HIERARCHICAL CLASSIFICATION OF THE LION

On moving down the hierarchy of groups, note that there are …

MORE SIMILARITIES
and
FEWER DIFFERENCES
between the members

Group	Other groups	Features
KINGDOM (Animalia)		All animals are ingestive heterotrophs
PHYLUM (Chordata)	Other phyla	All chordates have a notochord (→backbone)
CLASS (Mammalia)	Other classes	All mammals have fur and mammary glands
ORDER (Carnivora)	Other orders	All carnivores have well-developed carnassial (flesh-cutting) teeth
FAMILY (Felidae)	Other families	All felidae have retractable claws
GENUS (*Panthera*)	Other genera	All *Panthera* (big cats) can roar but cannot purr
SPECIES *Panthera leo*	Other species	All lions can mate and produce fertile offspring with other lions

DNA and chromosomes

Because of a range of physical forces this ladder-like molecule 'twists' itself into a **double helix**.

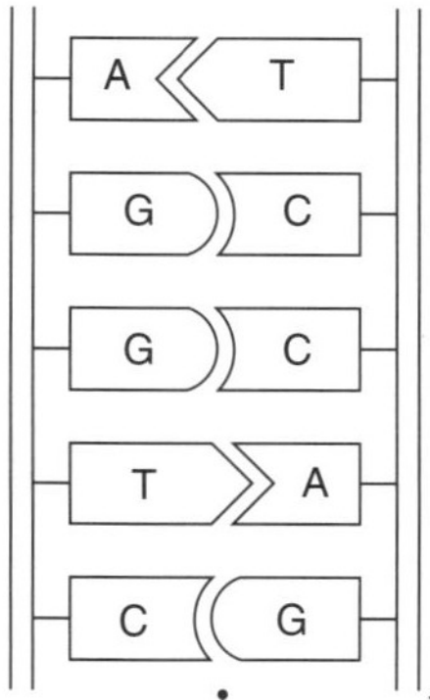

DNA is composed of chains of molecules called **nucleotide bases**. To protect the coded information from damage the bases (represented here by the code letters A for Adenine, G for Guanine, C for Cytosine and T for Thymine) form up into double chains.

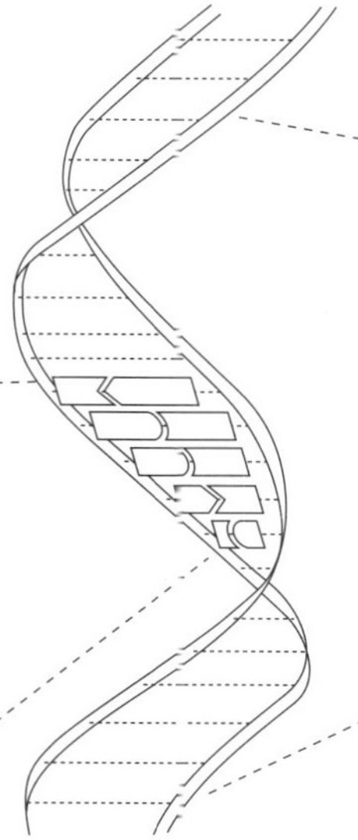

Very long DNA double helices wrap around a 'scaffold' made of proteins to form a **chromosome**.

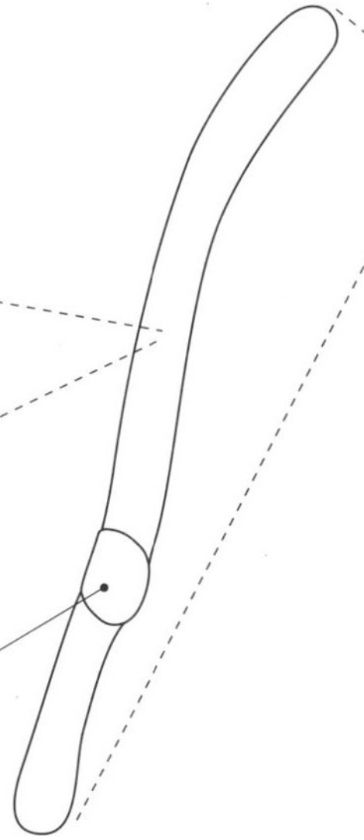

Centromere for attachment of spindle fibres during cell division.

The nucleus of each human cell (mature red blood cells excepted) contains 46 chromosomes arranged as **23 homologous pairs**.

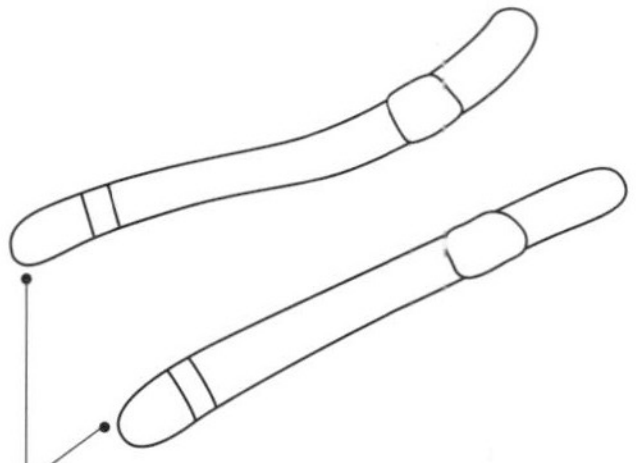

Homologous pairs of chromosomes carry the same genes (e.g. the gene for eye colour) but may carry alternative forms, or **alleles** of the same gene (e.g. blue on one chromosome, brown on the other)

The complete set of homologous chromosomes carries all of the genetic material for the individual = **genotype**.

This genotype is largely responsible for all of the physical measurable characteristics of this individual (for example, hair and eye colour) = **phenotype**.

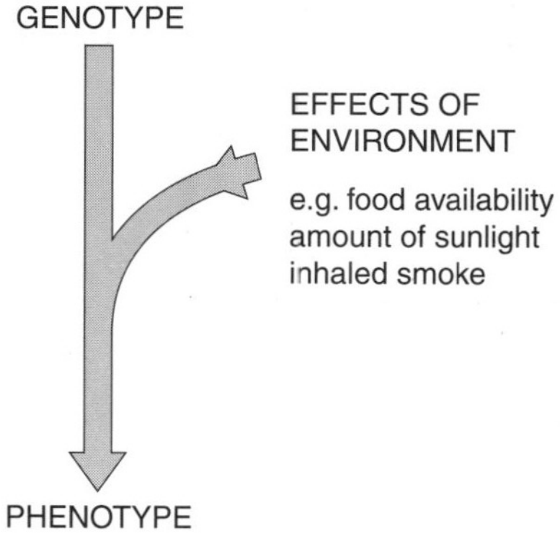

ALTERATIONS IN PHENOTYPE MAY RESULT FROM CHROMOSOME MUTATION

A mutation is an alteration in the DNA content of a cell. Chromosome mutations occur when the processes of cell division fail to work with complete accuracy. For example, during meiosis

WHAT SHOULD HAPPEN:

Adult 2n = 46 → Gametes n = 23, n = 23

BUT WHAT SOMETIMES HAPPENS:

Adult 2n = 46 → n = 24, n = 22

IF THESE GAMETES FUSE AT FERTILISATION (n = 23 + n = 24)

→ 2n = 47 (one chromosome too many)

In **Down's Syndrome** there is an 'extra' copy of the 21st chromosome → several phenotypic changes

e.g. narrowed eyes
broad forehead
heart abnormalities

DNA, genes and proteins

In the DNA molecule each 'code letter' (nucleotide base) **always** pairs with a specific partner i.e.
A·········T
and G·········C

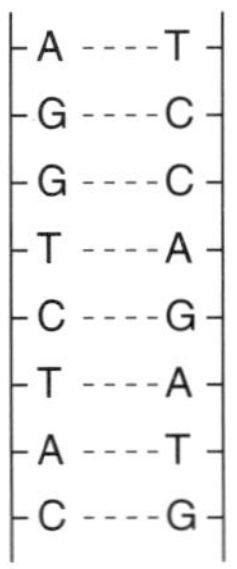

This is of great significance in explaining how the DNA molecule is copied during cell division.

The **genetic information in DNA** is carried as a sequence of 'codewords'. Each 'codeword' on the DNA is made up of **three** bases and each 'codeword' corresponds to a **single amino acid in a protein**.

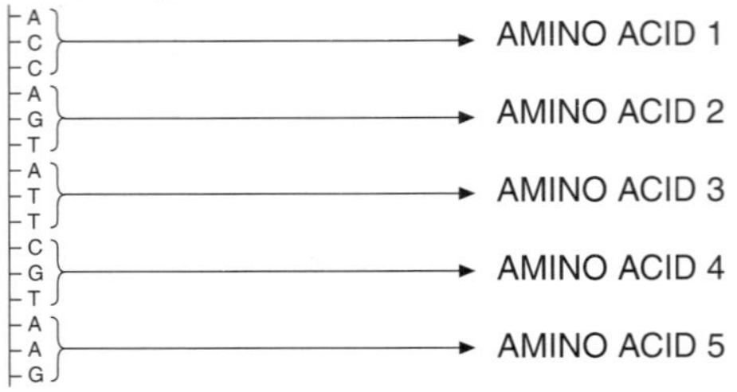

BUT DNA IS CONFINED TO THE NUCLEUS AND PROTEINS ARE MADE ON RIBOSOMES IN THE CYTOPLASM

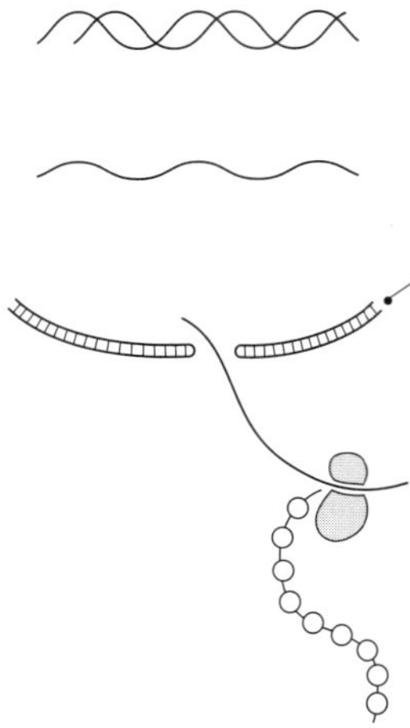

DNA on **chromosomes** in **nucleus**

One DNA strand is **transcribed** onto a messenger molecule

membrane around nucleus

messenger molecule leaves the nucleus

ribosome in cytoplasm **translates** message into sequence of amino acids i.e. into **protein**

Proteins are responsible for characteristics (e.g. haemoglobin is needed for oxygen transport)

Summary:
GENES → 'MESSENGER' → PROTEIN → CHARACTERISTICS

DNA REPLICATION

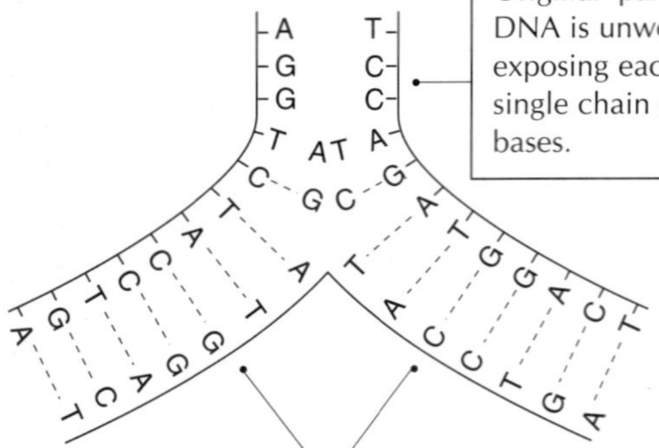

Original 'parent' DNA is unwound exposing each single chain of bases.

New DNA strands are synthesised using the unwound DNA as a template. Because of the specific base pairing the new strand will automatically have its base sequence determined: one DNA double helix becomes two identical double helices.

ALTERATIONS IN PHENOTYPE MAY RESULT FROM GENE MUTATIONS:

gene mutations occur when some part of the 'code letter' sequence in the DNA is altered. As a result a defective protein, or no protein at all, may be made.

DNA strand in chromosome

Gene - section of DNA which codes for 'normal' protein → NORMAL CHARACTERISTICS and NORMAL PHENOTYPE

Mutant gene - may be only a single 'letter' (base) which is altered → FAULTY PROTEIN → ALTERED PHENOTYPE

may be harmful, but could be beneficial under some conditions

- Sickle cell anaemia - faulty haemoglobin (poor oxygen transport **but** protection from malaria)
- Albinism - cannot make melanin pigment (danger of skin damage **but** may have better production of vitamin D)

RADIATION CAN INCREASE MUTATION RATES

Mutations **occur** spontaneously but there are factors which can increase the **rate** of mutation. Any factor which increases the rate of mutation is called a **mutagen**

- some chemicals — e.g. agent orange (defoliant in Vietnam)
- radiation — e.g. ultraviolet rays in sunlight, nuclear fall out

Cystic fibrosis in humans is an example of monohybrid inheritance

Cystic fibrosis results from an imbalance of chloride ions across the membranes of cells lining some of the major passageways of the body, causing dangerous accumulation of mucus.

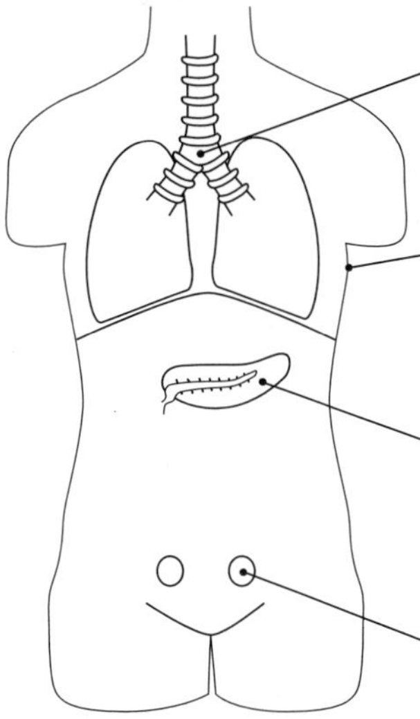

Lungs and airways are congested and readily infected.

Skin secretes an extremely salty sweat - diagnostic in even mild CF cases.

Pancreatic duct is congested, there is poor release of enzymes so that digestion is limited.

Gamete production is affected - CF adults are usually sterile or of very low fertility.

It is the most common inherited fatal disease - about 1 in 20 white Europeans is a carrier of the mutant allele.

If one parent's genotype is homozygous normal, the other parent's is homozygous mutant, then theoretically:

The homozygous normal parent is represented as NN, the parent with homozygous mutant genotype as nn, since in this case normal is dominant to mutant.

N N × n n — PARENTAL GENERATION

At meiosis, only one of the two chromosomes (thus only one of two alleles) can be transmitted to the gamete: Mendel's 1st Law.

N n — GAMETES

At fertilisation, fusion of gametes to form a zygote restores the diploid number.

1st FILIAL GENERATION

Nn

The allele which 'shows up' in this heterozygote is DOMINANT, the other allele (which remains 'hidden') is RECESSIVE

This individual is **genotypically** heterozygous, but **phenotypically** normal e.g. is a **carrier** of the mutant allele for cystic fibrosis.

If both parents are carriers (i.e. heterozygous)

PARENTAL GENERATION: Nn × Nn

GAMETES: N n N n

1st FILIAL (F1) GENERATION: NN Nn Nn nn

NN — Phenotype: no cystic fibrosis Genotype: normal

Nn, Nn — Phenotype: no cystic fibrosis Genotype: carrier of mutant allele

nn — Phenotype: has cystic fibrosis Genotype: homozygous for mutant allele

3 no cystic fibrosis : 1 with cystic fibrosis

A Punnett square can be used to predict the possible combinations of alleles in the zygote

Gametes from father / Gametes from mother	♂ N	n
♀ N	NN	Nn
n	Nn	nn

NB The 3 : 1 ratio is only approximate unless the number of offspring is very large (unlikely in humans), because

1. alleles may not be distributed between viable gametes in equal numbers
2. fusion of gametes is completely random. It is a matter of chance whether one male gamete fuses with a particular female gamete.

Huntington's Chorea is another example of monohybrid inheritance

- Sufferers from this condition usually show no symptoms until middle age (by which time they may have already passed on the mutant allele to their offspring) but then show a rapid, progressive deterioration of the nervous system and loss of co-ordination.
- The mutant allele is **dominant** so that heterozygotes suffer from Huntington's Chorea.
- A cross between two heterozygous parents theoretically produces offspring in the ratio
  3 sufferers : 1 normal
- Individuals with the mutant allele can be detected by sophisticated **gene probes**.

Sex linkage and the inheritance of sex

A KARYOTYPE is obtained by rearranging photographs of stained chromosomes observed during mitosis. Such a karyotype indicates that

① the chromosomes are arranged in **homologous pairs**. In humans there are 23 pairs and we say that the **diploid number** is 46 (2n = 46 = 2 x 23).

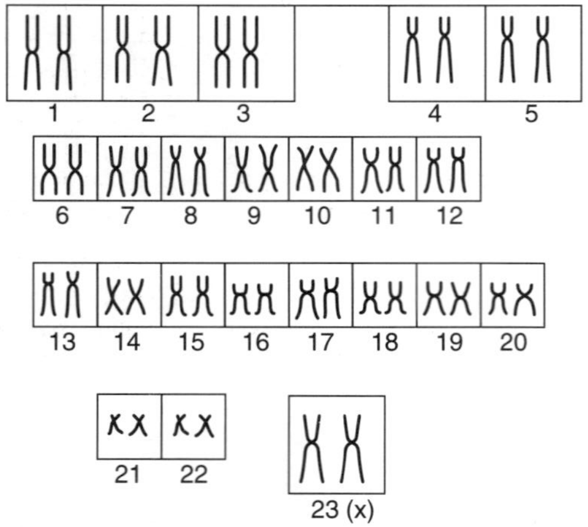

② whereas females have 22 pairs + XX in the karyotype, males have 22 pairs + XY i.e. the 23rd 'pair' would **not** be two copies of the X chromosome.

The Y chromosome is so small that there is little room for any genes other than those responsible for 'maleness', but the X chromosome can carry some genes as well as those for 'femaleness' - these additional genes are **X-linked** (usually described as **sex-linked**).

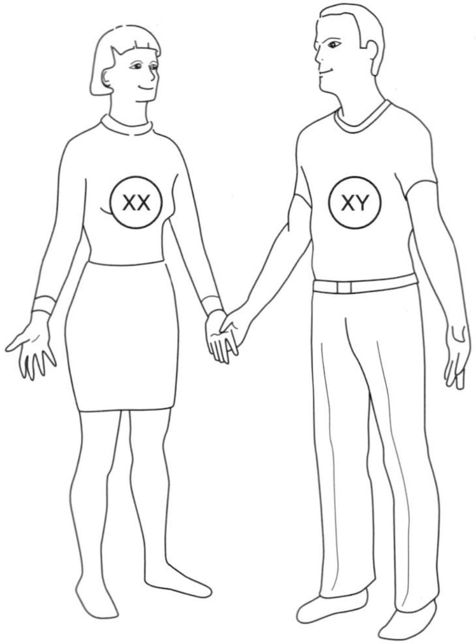

INHERITANCE OF SEX is a special form of monohybrid inheritance

	♂ (XY)	♀ (XX)
GAMETES	X Y	X X

F_1 generation: sex of offspring can be determined from a Punnett square

♀ GAMETES \ ♂ GAMETES	X	Y
X	XX (female)	XY (male)
X	XX (female)	XY (male)

- theoretically there should be 1 : 1 ratio of male : female
- the male gamete determines the sex of the offspring

SEX LINKAGE: Genes on the X chromosome can be inherited by either sex, but the male can only receive **one** of the possible alleles as he **must** be XY. The female will be XX and thus may be **either homozygous or heterozygous** for the X-linked allele.

For example, **haemophilia** (in which blood fails to clot properly) is an X-linked condition.

Normal allele = H, mutant allele = h

PARENTS	X^HX^h	×	X^HY
	female, normal phenotype but a carrier of the mutant allele		male, normal phenotype
GAMETES	X^H X^h		X^H Y

F_1 OFFSPRING

♀ GAMETES \ ♂ GAMETES	X^H	Y
X^H	X^HX^H	X^HY
X^h	X^HX^h	X^hY

i.e. two 'normal' parents can produce a haemophiliac son

N.B. A haemophiliac daughter (X^hX^h) could only be produced if **both** parents contributed X^h i.e. if the father was haemophiliac, in which case his condition would be known and the daughter's condition might be expected.

Another important X-linked condition is **red-green colour blindness.**
Many more males than females cannot distinguish red from green

Cystic fibrosis in humans is an example of monohybrid inheritance

Cystic fibrosis results from an imbalance of chloride ions across the membranes of cells lining some of the major passageways of the body, causing dangerous accumulation of mucus.

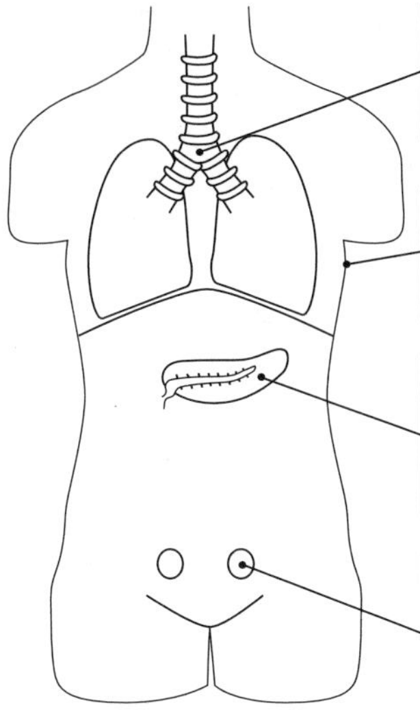

Lungs and airways are congested and readily infected.

Skin secretes an extremely salty sweat - diagnostic in even mild CF cases.

Pancreatic duct is congested, there is poor release of enzymes so that digestion is limited.

Gamete production is affected - CF adults are usually sterile or of very low fertility.

It is the most common inherited fatal disease - about 1 in 20 white Europeans is a carrier of the mutant allele.

If one parent's genotype is homozygous normal, the other parent's is homozygous mutant, then theoretically:

The homozygous normal parent is represented as NN, the parent with homozygous mutant genotype as nn, since in this case normal is dominant to mutant.

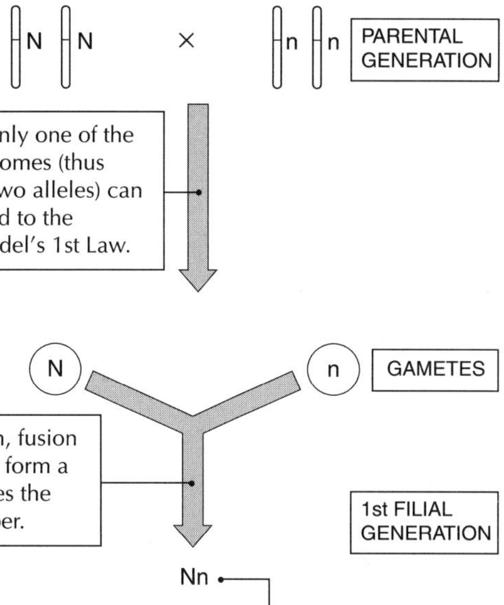

At meiosis, only one of the two chromosomes (thus only one of two alleles) can be transmitted to the gamete: Mendel's 1st Law.

At fertilisation, fusion of gametes to form a zygote restores the diploid number.

The allele which 'shows up' in this heterozygote is DOMINANT, the other allele (which remains 'hidden') is RECESSIVE

This individual is **genotypically** heterozygous, but **phenotypically** normal e.g. is a **carrier** of the mutant allele for cystic fibrosis.

If both parents are carriers (i.e. heterozygous)

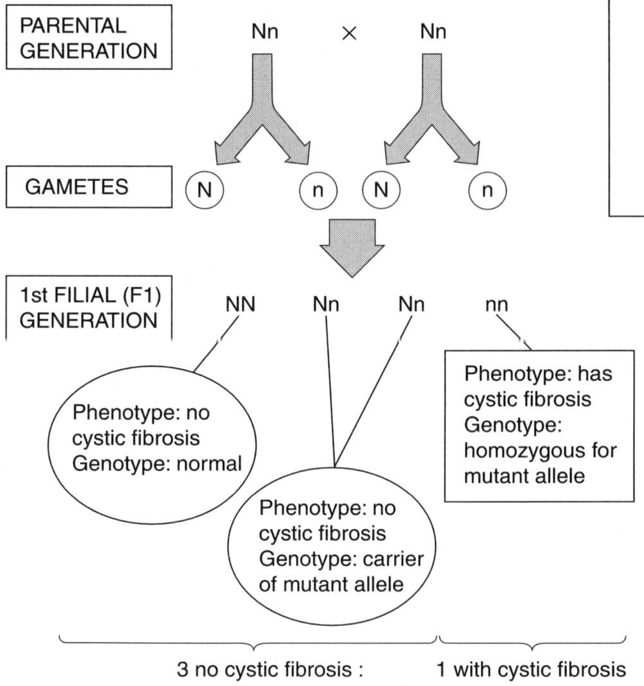

A Punnett square can be used to predict the possible combinations of alleles in the zygote

Gametes from mother \ Gametes from father	♂ N	n
♀ N	NN	Nn
n	Nn	nn

NB The 3 : 1 ratio is only approximate unless the number of offspring is very large (unlikely in humans), because
1. alleles may not be distributed between viable gametes in equal numbers
2. fusion of gametes is completely random. It is a matter of chance whether one male gamete fuses with a particular female gamete.

Huntington's Chorea is another example of monohybrid inheritance
- Sufferers from this condition usually show no symptoms until middle age (by which time they may have already passed on the mutant allele to their offspring) but then show a rapid, progressive deterioration of the nervous system and loss of co-ordination.
- The mutant allele is **dominant** so that heterozygotes suffer from Huntington's Chorea.
- A cross between two heterozygous parents theoretically produces offspring in the ratio
  3 sufferers : 1 normal
- Individuals with the mutant allele can be detected by sophisticated **gene probes**.

Sex linkage and the inheritance of sex

A KARYOTYPE is obtained by rearranging photographs of stained chromosomes observed during mitosis. Such a karyotype indicates that

① the chromosomes are arranged in **homologous pairs**. In humans there are 23 pairs and we say that the **diploid number** is 46 (2n = 46 = 2 x 23).

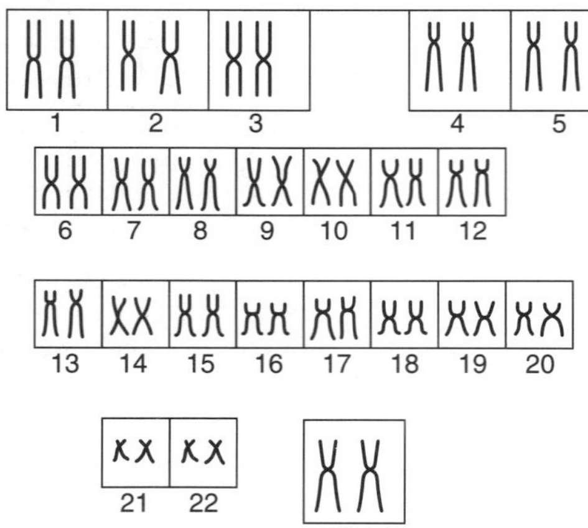

② whereas females have 22 pairs + XX in the karyotype, males have 22 pairs + XY i.e. the 23rd 'pair' would **not** be two copies of the X chromosome.

The Y chromosome is so small that there is little room for any genes other than those responsible for 'maleness', but the X chromosome can carry some genes as well as those for 'femaleness' - these additional genes are **X-linked** (usually described as **sex-linked**).

INHERITANCE OF SEX is a special form of monohybrid inheritance

	♂ (XY)	♀ (XX)
GAMETES	(X) (Y)	(X) (X)

F_1 generation: sex of offspring can be determined from a Punnett square

♀ GAMETES \ ♂ GAMETES	X	Y
X	XX (female)	XY (male)
X	XX (female)	XY (male)

- theoretically there should be 1 : 1 ratio of male : female
- the male gamete determines the sex of the offspring

SEX LINKAGE: Genes on the X chromosome can be inherited by either sex, but the male can only receive **one** of the possible alleles as he **must** be XY. The female will be XX and thus may be **either homozygous or heterozygous** for the X-linked allele.

For example, **haemophilia** (in which blood fails to clot properly) is an X-linked condition.

Normal allele = H, mutant allele = h

PARENTS	X^HX^h	×	X^HY
	female, normal phenotype but a carrier of the mutant allele		male, normal phenotype

GAMETES (X^H) (X^h) (X^H) (Y)

F_1 OFFSPRING

♀ GAMETES \ ♂ GAMETES	X^H	Y
X^H	X^HX^H	X^HY
X^h	X^HX^h	X^hY

i.e. two 'normal' parents can produce a haemophiliac son

N.B. A haemophiliac daughter (X^hX^h) could only be produced if **both** parents contributed X^h i.e. if the father was haemophiliac, in which case his condition would be known and the daughter's condition might be expected.

Another important X-linked condition is **red-green colour blindness.**
Many more males than females cannot distinguish red from green

Variation and natural selection may lead to evolution of species

Environmental resistance describes the factors, e.g. food availability, which limit the growth of populations. Animals and plants which are best **adapted** to their environment suffer from less environmental resistance. **Adaptations** are often structural, but they can also be biochemical or behavioural.

These adaptations arise because there is **variation** between the different members of a population. Charles Darwin studied many examples of these adaptations, and published his conclusions in *The Origin of Species by means of Natural Selection*. His observations can be summarised under a number of headings.

Over-production: All organisms produce more offspring than can possibly survive, and yet populations remain relatively stable.

e.g. a female peppered moth may lay 500 eggs, but the moth population does not increase by 25000%!

Struggle for existence: Organisms experience environmental resistance i.e. they compete for the limited resources within the environment.

e.g. several moths may try to feed on the same nectar-producing flower.

Variation: Within the population there may be some characteristics which make the organisms which possess them more suited for this severe competition.

e.g. moths might be stronger fliers, have better feeding mouthparts, be better camouflaged while resting or be less affected by rain.

Survival of the fittest: Individuals which are most successful in the struggle for existence (i.e. which are the best suited/adapted to their environment) will survive more easily than those without these advantages.

e.g. peppered moths: dark coloured moths resting on soot-covered tree trunks will be less likely to be captured by predators.

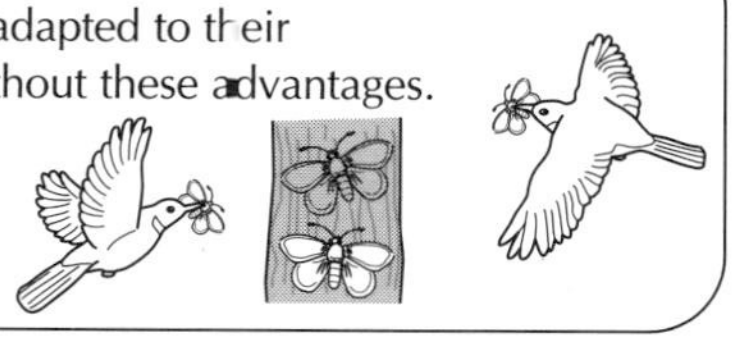

Advantageous characteristics are passed on to offspring: The well-adapted individuals breed more successfully than those which are less well-adapted - they pass on their genes to the next generation. This process is called **natural selection**.

e.g. dark coloured moth parents will produce dark coloured moth offspring.

HOW GENETIC VARIATION MAY ARISE

- both chromosomal and gene **mutation** provide **raw material for variation**
- meiosis and sexual reproduction then **re-assort this raw material** to provide **many new genotypes**.

① ② MEIOSIS: CROSSING OVER and INDEPENDENT ASSORTMENT

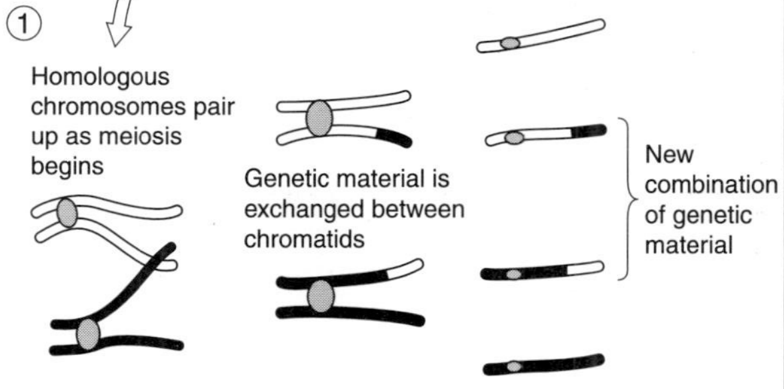

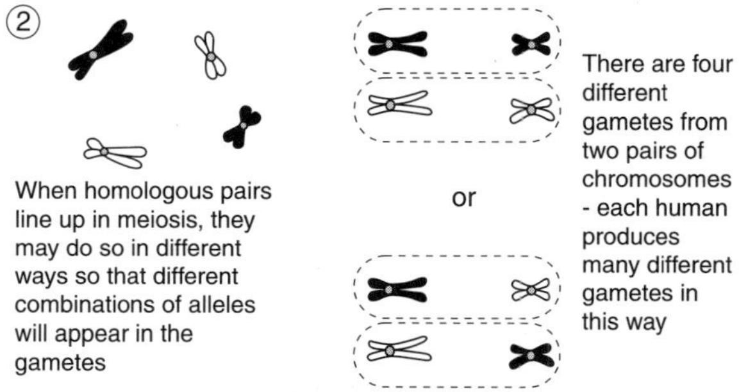

③ FERTILISATION

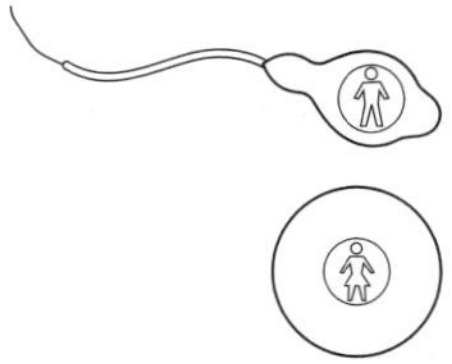

Any male gamete (300 000 000 different in each ejaculation!)

can combine with

any female gamete (every one is likely to be different)

MUTATION + ① + ② + ③ ⟹ VIRTUALLY LIMITLESS VARIATION!

Summary

VARIATION —NATURAL SELECTION→ ADAPTATION —CONTINUED NATURAL SELECTION→ HIGHLY-ADAPTED POPULATIONS = NEW SPECIES

Genetic engineering (recombinant DNA technology)

depends on enzymes and culture of microorganisms

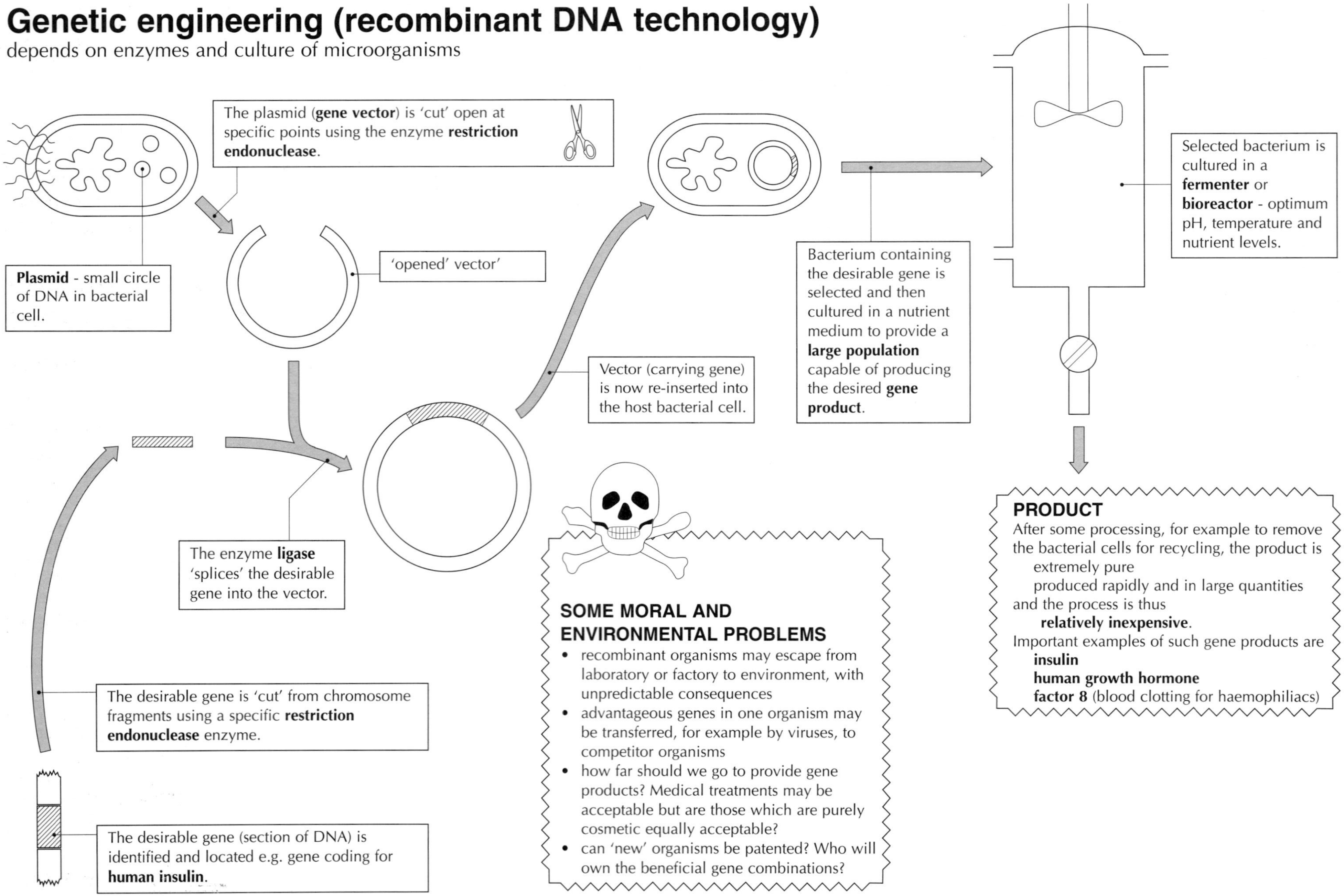

SOME MORAL AND ENVIRONMENTAL PROBLEMS

- recombinant organisms may escape from laboratory or factory to environment, with unpredictable consequences
- advantageous genes in one organism may be transferred, for example by viruses, to competitor organisms
- how far should we go to provide gene products? Medical treatments may be acceptable but are those which are purely cosmetic equally acceptable?
- can 'new' organisms be patented? Who will own the beneficial gene combinations?

PRODUCT

After some processing, for example to remove the bacterial cells for recycling, the product is
extremely pure
produced rapidly and in large quantities
and the process is thus
relatively inexpensive.
Important examples of such gene products are
insulin
human growth hormone
factor 8 (blood clotting for haemophiliacs)

Gene transfer can promote desirable characteristics

RESISTANCE TO PESTS AND HERBICIDES

A gene is inserted into the plant which enables it to make **insecticidal crystal protein** (I.C.P.) which affects the gut of the caterpillars so that they cannot feed and eventually die.

The crop plant has a gene transferred into it which makes it resistant to herbicides. The field of growing crop can then be sprayed with the herbicide which will selectively kill the 'weeds' since they do not possess the 'resistance' gene.

TRANSFERRING GENES WITH *Agrobacterium tumefaciens*

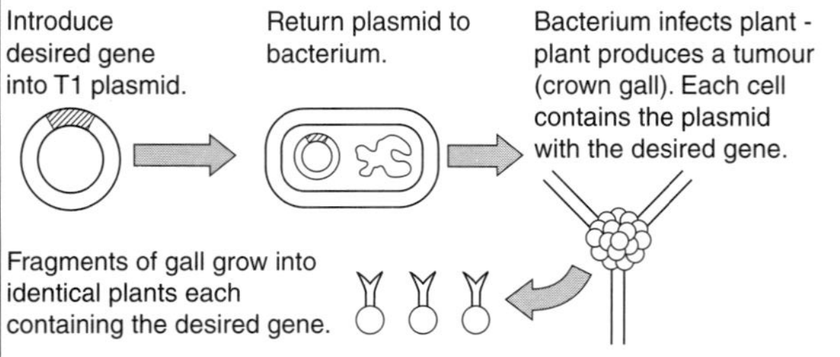

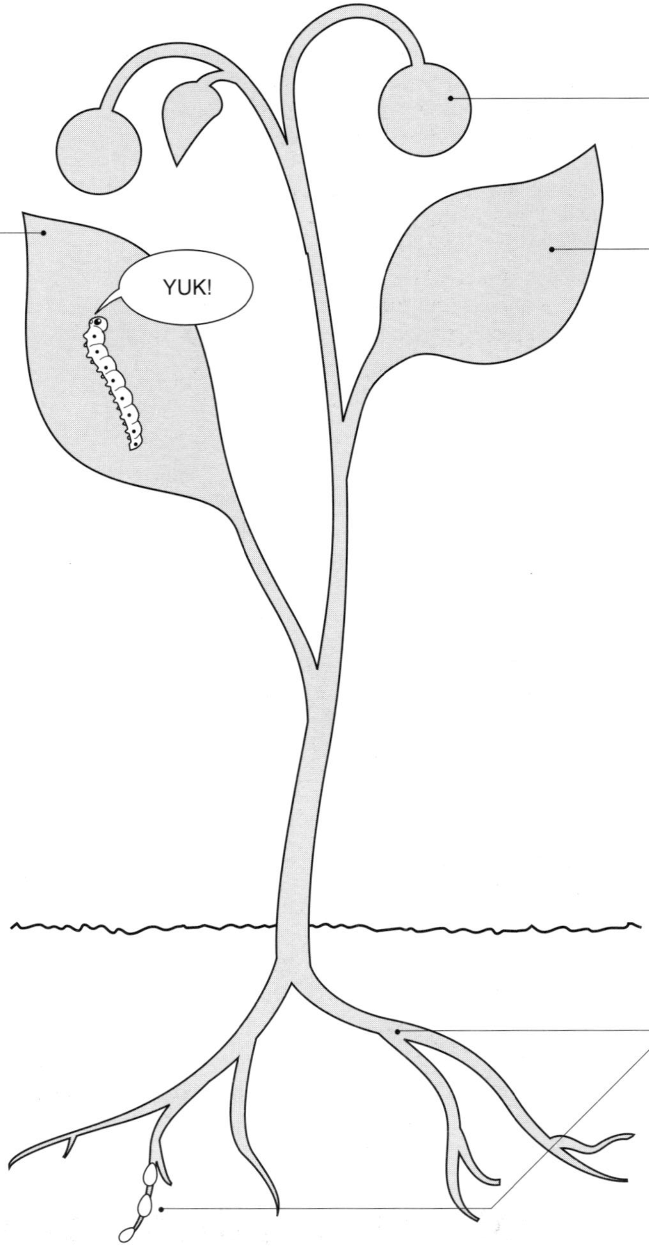

Improved shelf life: many fruits are wasted because they deteriorate before they can be sold/eaten. A gene has been introduced into tomatoes which inhibits the enzymes causing deterioration - they new '**flavr-savr**' tomatoes last for several weeks.

Resistance to environmental conditions: the use of crop plants for food is limited by the conditions in which the plants will grow successfully. Gene transfer has produced

- **drought resistance** - plants with thicker, more waxy cuticles which can grow well in more arid areas.
- **frost resistance** - plants damaged by frost are usually infected by a bacterium *Pseudomonas syringae* which produces an 'ice-protein'. A bacterium engineered to be unable to make the 'ice-protein' is sprayed onto strawberry plants and competes with the damage-causing strain.
- **uniform fruiting** - plants produce flowers and fruits in response to daylight. Soya beans plants have been engineered to produce beans even in temperate regions of the world with a different light regime to their normal habitat.
- **wind damage** - soya plants have been engineered to have **stronger** stems of a **more uniform height** which makes them more resistant to wind damage and makes machine harvesting of the beans more efficient/economical.

Nitrogen fixation: this process involves the reduction of nitrogen gas from the atmosphere into a form suitable for conversion into amino acids, nucleotides and other essential compounds.

N_2 (from atmosphere) → NH_4^+ (ammonium ions) → amino acids and other compounds

this key step is controlled by enzymes coded for by 12 genes called the Nif genes.

Most plants cannot fix nitrogen but it is hoped that gene transfer might either

- insert these genes directly into a plant
- make a plant more susceptible to the formation of root nodules with nitrogen-fixing bacteria of the genus *Rhizobium*.

This could

- produce cereal crops which also manufacture large amounts of protein;
- reduce the demand for nitrogenous fertilisers.

GENE TRANSFER IS ALSO IMPORTANT IN ANIMALS

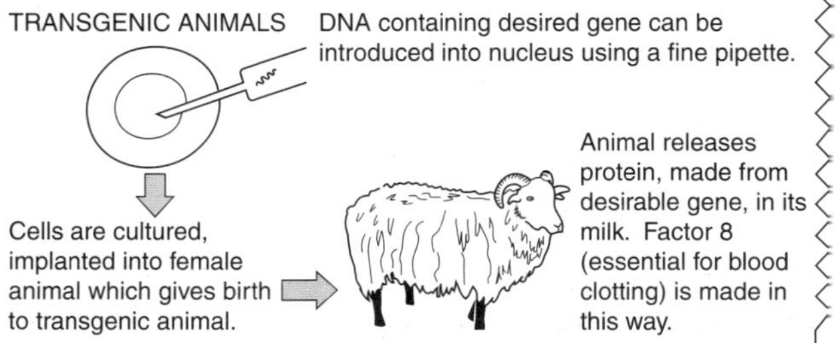

Cloning - the production of genetically identical individuals - has many applications.

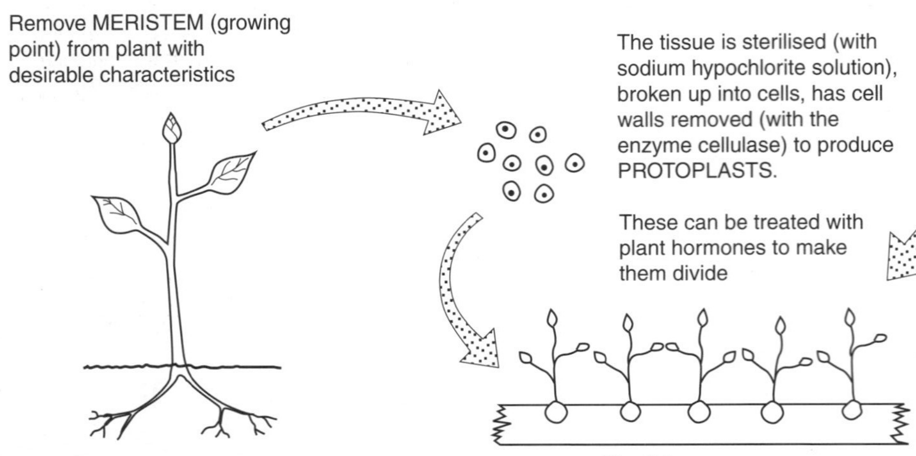

TISSUE CLONING: allows growth of many identical individuals in a small space - ideal for transport, e.g. rubber 'trees' can be started in Britain then transferred to Malaysia for planting.

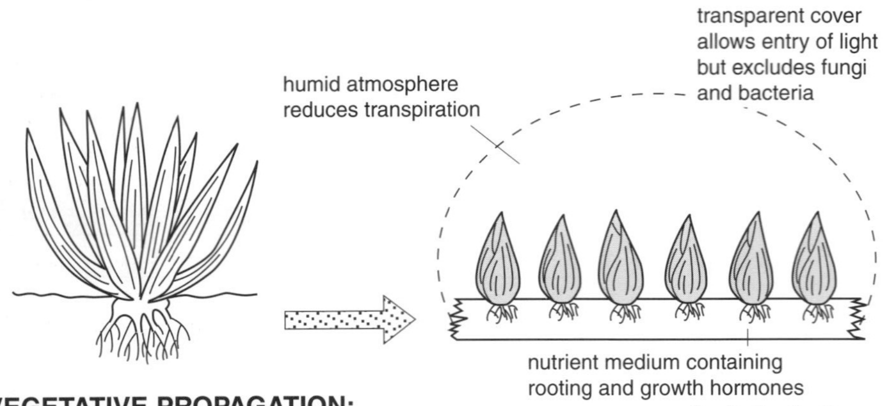

VEGETATIVE PROPAGATION:
Fragments of plants can be separated from the parent and induced to grow in a propagator. These fragments, e.g. root or leaf cuttings, are genetically identical and if kept under ideal conditions will grow by mitosis into plants **identical to each other and to the parent**.

Many ornamental house plants are produced in this way.

'IN VITRO' (TEST TUBE) FERTILISATION

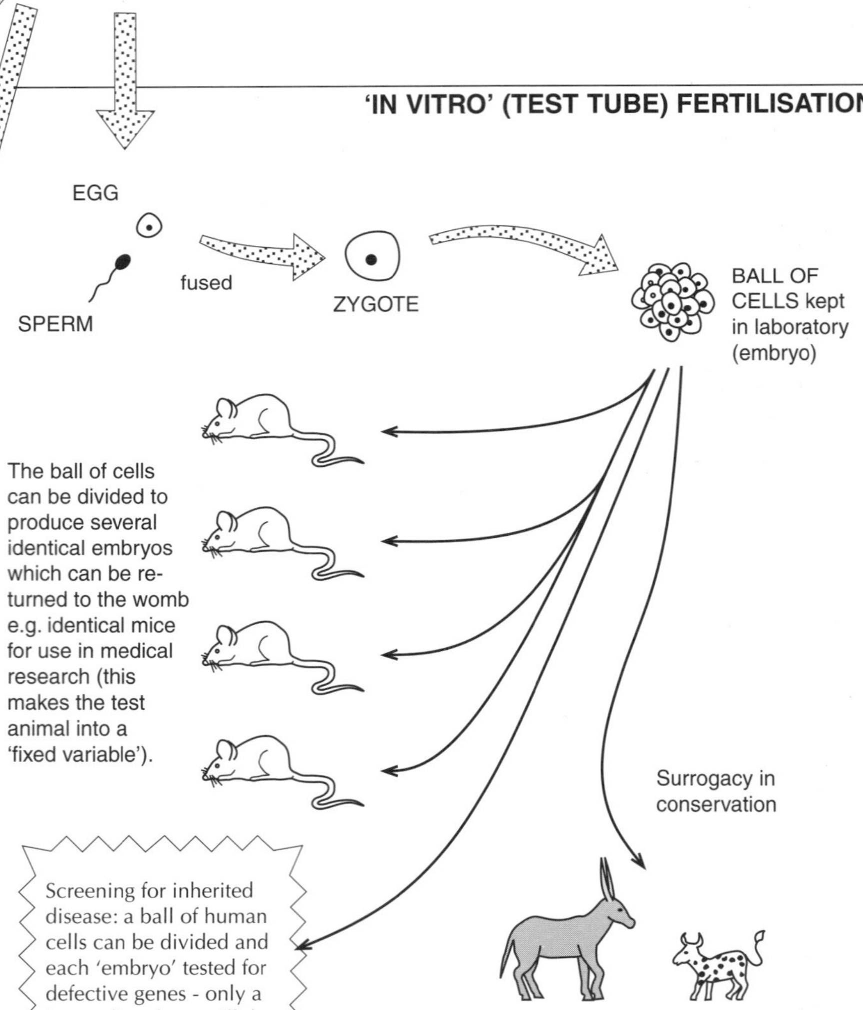

The ball of cells can be divided to produce several identical embryos which can be returned to the womb e.g. identical mice for use in medical research (this makes the test animal into a 'fixed variable').

Screening for inherited disease: a ball of human cells can be divided and each 'embryo' tested for defective genes - only a 'normal' embryo will then be returned to the mother's womb to develop into a foetus.

•zygotes of uncommon species (e.g. the bongo) can be placed in the womb of a common species (e.g. eland) - in this way the maximum number of 'rare' offspring can be produced.

Bioreactors/fermenters exploit microbes for commercial reasons.

WHY USE MICRO-ORGANISMS?

- carry out reactions at moderate temperatures
  → **great savings on fuel**
- are very efficient
  → **less waste and purer products**
- can be 'genetically engineered' to produce **compounds needed for humans**.

Paddle stirrers: continuously mix the contents of the bioreactor
- ensure micro-organisms are always in contact with nutrients
- ensures an even temperature throughout the fermentation mixture
- for aerobic (oxygen-requiring) fermentations the mixing may be carried out by an **airstream**.

Input for microbial culture: the organisms which will carry out the fermentation process are cultured separately until they are growing well.

Nutrient input: the micro-organisms require
- an **energy source** - usually carbohydrate
- **growth materials** - amino acids, or ammonium salts which can be converted to amino acids, are required for protein synthesis.

Constant temperature water jacket: the temperature is controlled so that it is high enough to promote enzyme activity but not so high that enzymes and other proteins in the microbes are denatured.
- this initially involves **heating** to initiate the fermentation
- fermentation releases heat so that later stages may require **cooling**.

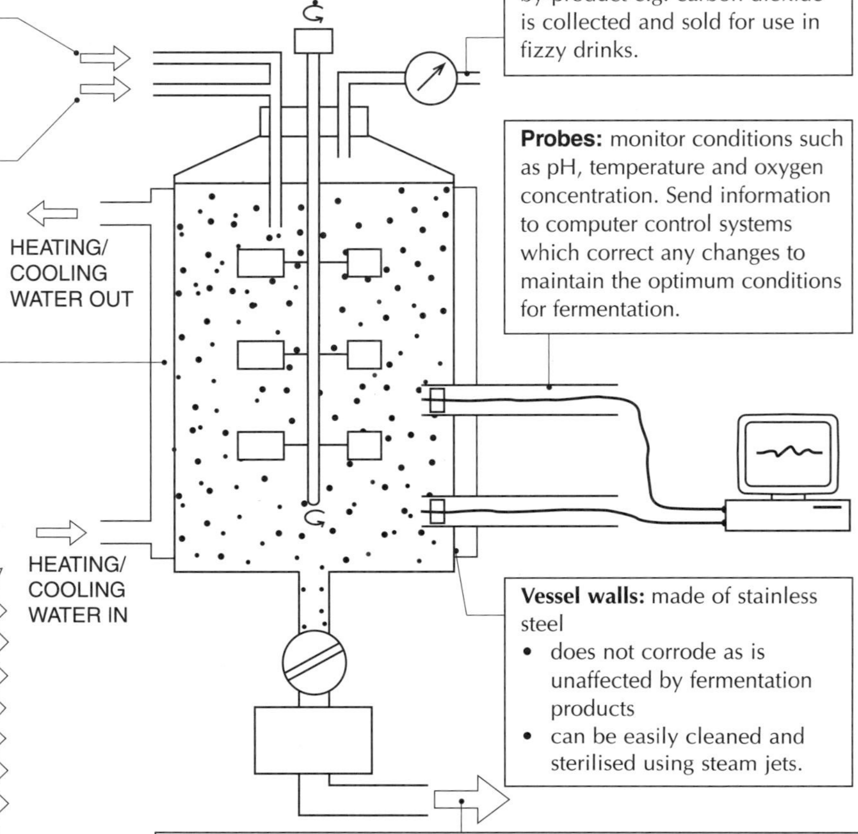

Gas outlet: gas may be evolved during fermentation. This must be released to avoid pressure damage, and may be a valuable by-product e.g. carbon dioxide is collected and sold for use in fizzy drinks.

Probes: monitor conditions such as pH, temperature and oxygen concentration. Send information to computer control systems which correct any changes to maintain the optimum conditions for fermentation.

Vessel walls: made of stainless steel
- does not corrode as is unaffected by fermentation products
- can be easily cleaned and sterilised using steam jets.

Sterile conditions are essential: the culture must be pure and all nutrients/equipment sterile to
- avoid competition for expensive nutrients
- limit the danger of disease-causing organisms contaminating the product.

WASTE NOT, WANT NOT!

- ideally the nutrients should be waste products of other reactions e.g. remains of crops grown for other purposes.
- the micro-organisms should be used to start or maintain new cultures.

Further processing of product may be necessary
- to separate the micro-organism from the desired product. In some fermentation systems these micro-organisms may then be returned to the vessel to continue the process.
- to prepare the product for sale or distribution - this often involves **drying** or **crystallisation**.

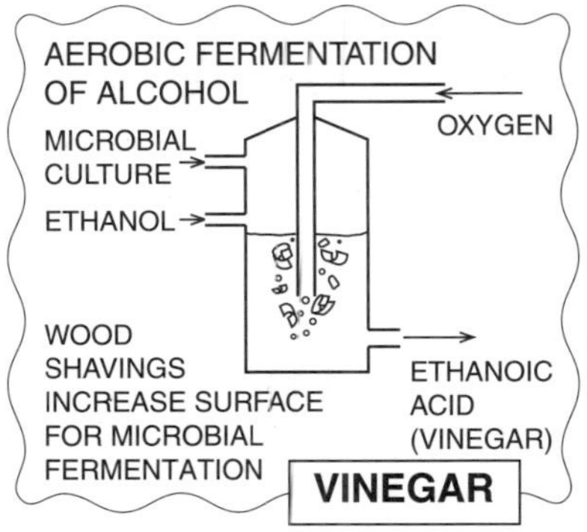

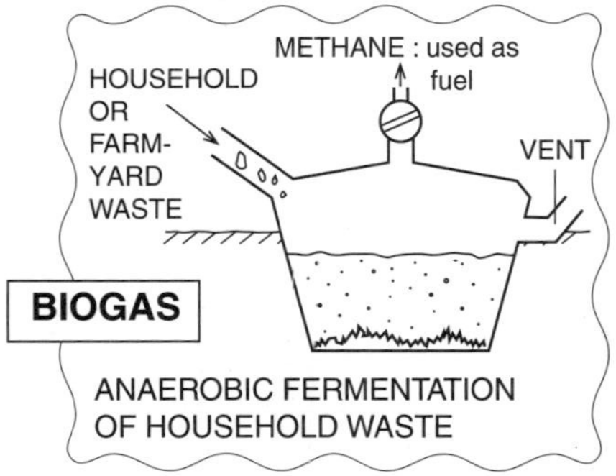

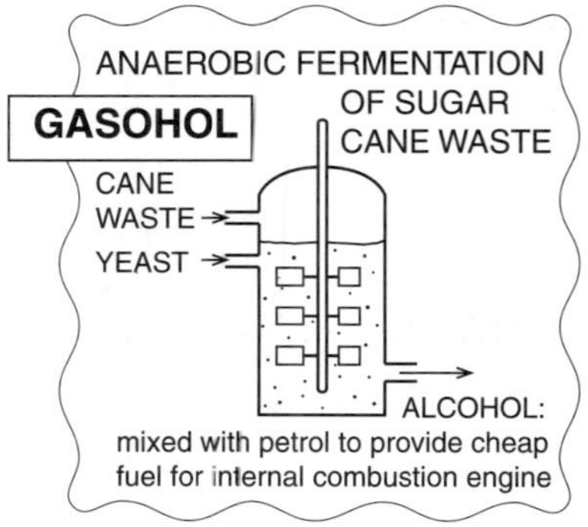

PENICILLIUM MOULD is a **saprotroph** which reproduces by **sporulation**.

Other, related topics can be found on pages ....

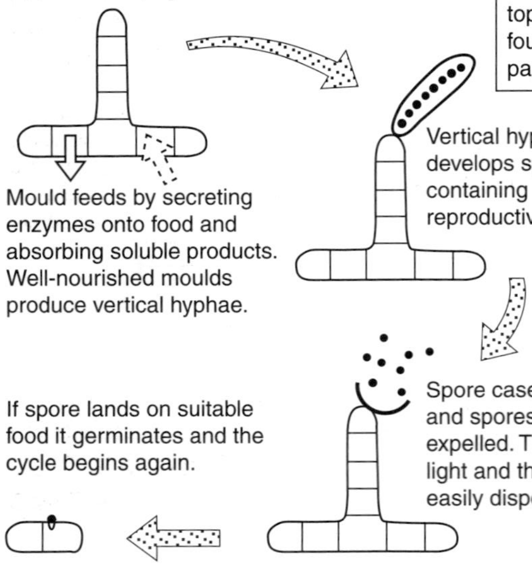

LARGE SCALE PRODUCTION OF PENICILLIN

is a good example of the commercial use of fermentation reactions.

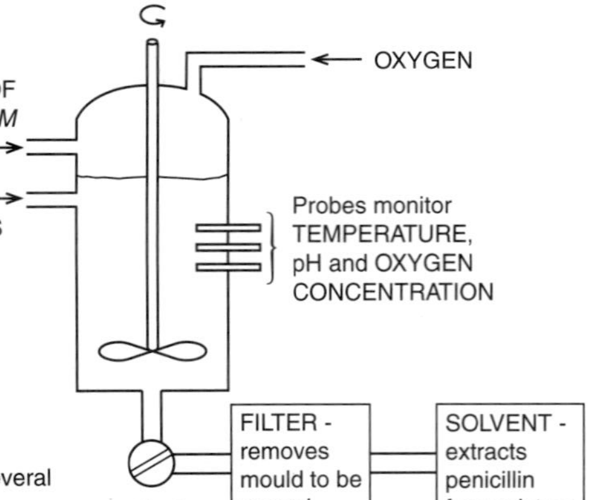

There are several *Penicillium* moulds which may be used to produce antibiotics. The most widely used is *Penicillium crysogemum*.

PENICILLIN IS AN ANTIBIOTIC

The *Penicillium* mould may make products which it secretes into its environment to kill off any disease-causing or competitive micro-organisms. **A compound made by a mould to kill off any other micro-organism is called an ANTIBIOTIC**.

Antibiotics are valuable in the control of bacterial diseases because they affect bacterial cells but rarely affect human or animal cells.

Some important diseases caused by bacteria and thus suitable for antibiotic treatment are:

- Dysentery, Food poisoning } affect gut
- Syphilis, Gonorrhoea } affect reproductive organs
- Pneumonia - affects lungs
- Botulism - affects nervous system

ACTION OF PENICILLIN

Bacteria multiply very rapidly by **binary fission** i.e. they grow and then divide into two. This can happen very rapidly (every 20 minutes), causing disease as the bacterial metabolism competes with or inhibits the activity of human cells.

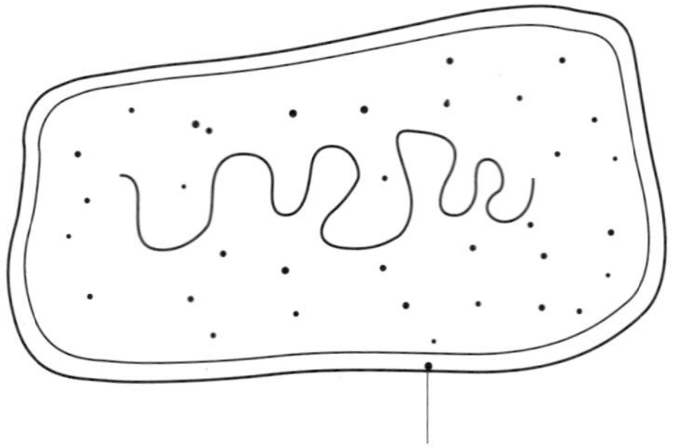

Bacterial cells are surrounded by a **cell wall**. **Penicillin prevents the bacterium from making components of the cell wall** - the bacterial cell is therefore weakened.

Antibiotics may be **bacteriocidal** (i.e. they kill the pathogenic bacterium directly) or they may be **bacteriostatic** (i.e. they prevent replication, leaving the host's defences to kill the existing pathogens).

Penicillin is bacteriocidal at high concentrations, but this may have some side-effects in humans and encourage development of strains of bacteria resistant to treatment by penicillin. **Penicillin is usually given at doses which are bacteriostatic.**

RESISTANT BACTERIAL STRAINS MAY DEVELOP

Bacteria breed rapidly and populations are enormous. Within these populations rare **mutations** may produce cells which are **antibiotic-resistant**. Over-use of antibiotics may destroy 'normal' bacterial cells but allow 'resistant' cells to survive. These may multiply to form a resistant poulation.

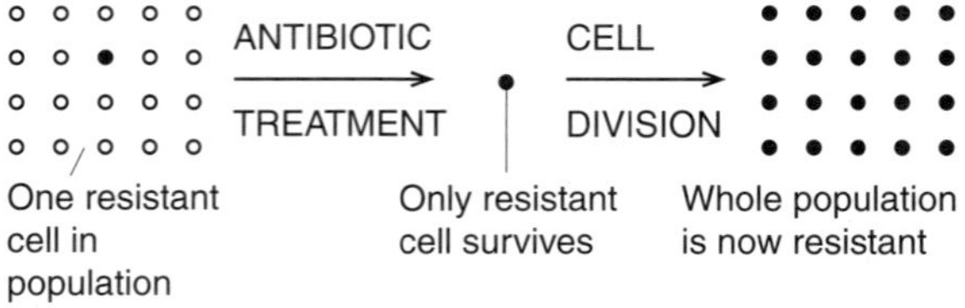

This is a form of **artificial selection**.

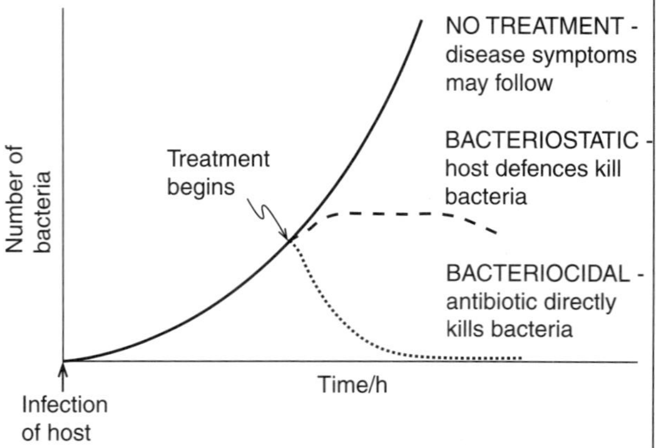

Bacteria and food production

Bacterial cells are surrounded by a cell wall. They have cytoplasm **but no nucleus**. Many of them are **heterotrophic** - they feed on organic foods.

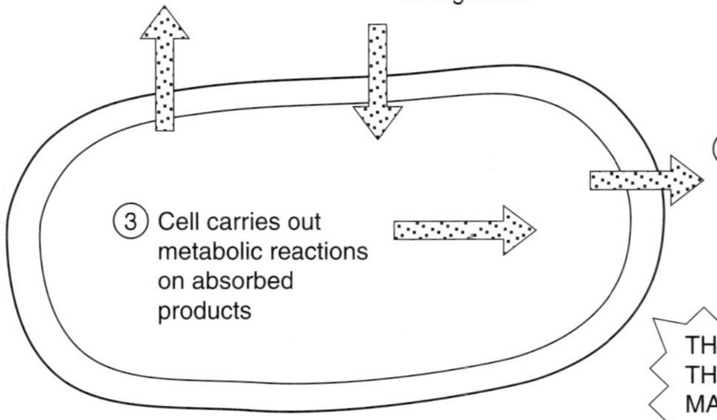

THE EFFECTS OF THESE PRODUCTS MAY BE OF GREAT IMPORTANCE TO HUMANS

SINGLE CELL PROTEIN (S.C.P.)

This is made up of the dried cells of many millions of bacteria, which are grown up in a **fermenter/bioreactor**.

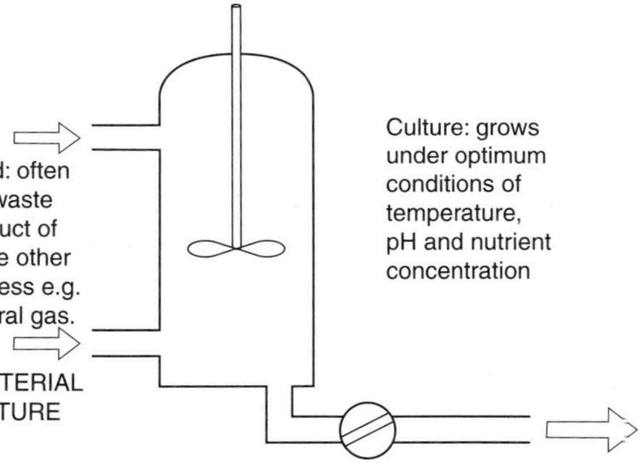

An important food for cattle (limited use for humans).

YOGHURT: this is milk which has been slightly 'soured' by the excretion of lactic acid from bacterial cultures.

HOMOGENISATION → 'THICK MILK' with a uniform distribution of fat globules

PASTEURISATION: heat at 90°C for 15-30 minutes, then cool to 45°C. → Milk which is free from dangerous bacteria - also thicker due to denaturation of casein (milk protein)

INCUBATION: the culture of bacteria (*Lactobacillus*) maintained at 45°C convert LACTOSE (MILK SUGAR) → LACTIC ACID → 'SOURED', partially clotted milk with mildly acidic taste: 'natural yoghurt'

COOLING TO 4°C halts reaction - fruit / nuts may be added →

Flavoured product for sale/consumption.

VINEGAR : is produced by the aerobic fermentation of alcohol

ALCOHOL + OXYGEN $\xrightarrow{\textit{Acetobacter}}$ ETHANOIC ACID (VINEGAR) + WATER

CHEESE : may require ripening with enzymes from bacteria.

MILK $\xrightarrow{\text{Rennin}}$ CURD $\xrightarrow{\text{Bacteria}}$ CHEESE 'flavoured' by fat/protein breakdown

FOOD SPOILAGE may be cause by bacterial enzymes and toxins. Many techniques for **food preservation** are designed to prevent bacterial growth.

Anaerobic respiration

involves incomplete breakdown of glucose.

GLUCOSE

PYRUVIC ACID - a molecule formed part way through the complete breakdown of glucose. **Complete breakdown** would require **oxygen**.

In animals (and some bacteria)

Only about 5% of the energy which would be released by aerobic (oxygen present) respiration, but enough to keep the muscles working for a short time

ENERGY + LACTIC ACID

This by-product is toxic (poisonous) and would quickly inhibit the muscles unless it is quickly removed.

Muscle aches and stiffness after exercise are partly the result of lactic acid which remains in the muscle because the warming-down period was not long enough to 'wash it out' of the muscle.

EXERCISE AND LACTIC ACID

Before exercise, or during mild exercise, the blood can deliver enough oxygen to keep aerobic respiration at a level high enough to supply the body's demands for oxygen.

During severe exercise oxygen delivery increases but not enough to maintain aerobic respiration. The body's demands for energy must now be partially met by anaerobic respiration. Lactic acid is 'washed' from the muscles into the blood.

During recovery lactic acid continues to be 'washed' from the muscles but is removed by oxidation in the liver. The additional oxygen needed for this makes up the oxygen debt built up during exercise.

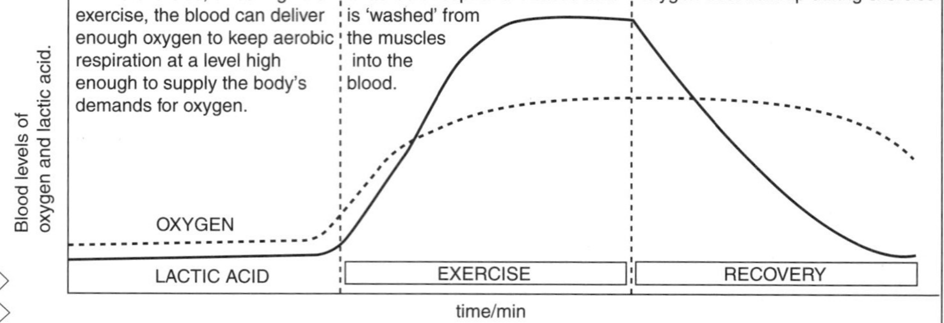

In plants, and, more importantly in yeast.

ENERGY + CARBON DIOXIDE + ETHANOL

This energy is required by the yeast in order to carry out the reactions required for cell growth and multiplication.

CARBON DIOXIDE IS PARTICULARLY IMPORTANT IN BREAD MAKING

Yeast + sugar warmed together = RAISING AGENT

FLOUR, SALT, WARM WATER → DOUGH → FERMENTATION → RISEN DOUGH → BAKING • kills yeast • evaporates alcohol → BREAD

(carbon dioxide 'bubbles' cause the dough to swell)

If a culture of yeast is supplied with glucose and water, at a temperature of around 28°C it will reproduce by **budding**.

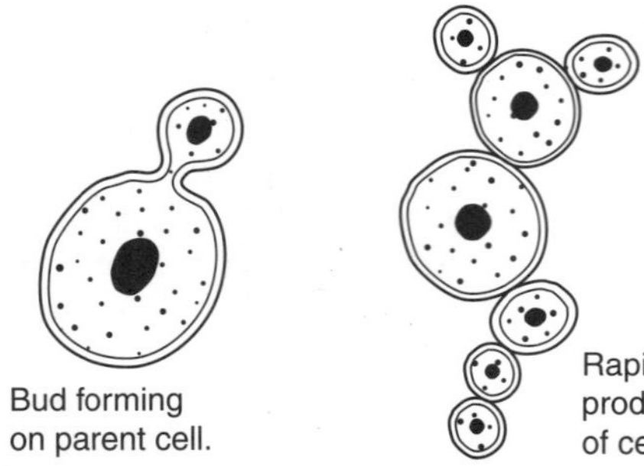

Bud forming on parent cell.

Rapid budding produces chains of cells.

NB Alcohol, like lactic acid, is a poison. It eventually kills the yeast cells which produce it, and does the same to human cells if taken in too large a quantity!

BREWING DEPENDS ON ALCOHOLIC FERMENTATION

Hops (for flavour)

Germinating barley (for malt sugar)

YEAST

28°C

Alcohol solution ('flat' since all carbon dioxide has escaped)

more sugar

stoppered bottles

Beer made 'fizzy' by carbon dioxide from second fermentation

Ecology is the study of living organisms in relation to their environment.

THIS MAY INVOLVE

ADAPTATIONS OF SINGLE ORGANISMS

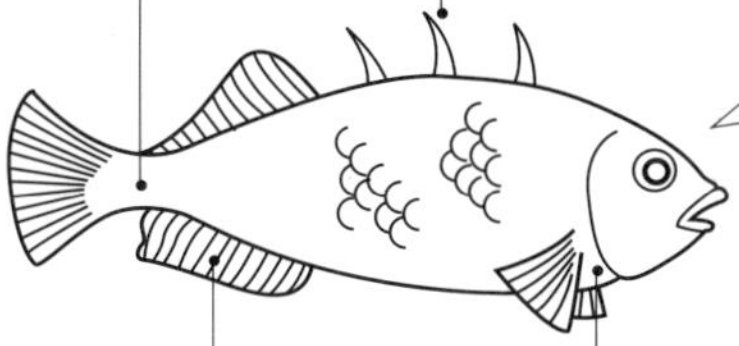

These structural adaptations help the individual organism to survive in its **habitat**.

HOW GROUPS OF ORGANISMS INTERACT TO FORM 'UNITS OF THE ENVIRONMENT'

Population - all the members of the same species within a given physical area e.g. all of the 3-spined stickle-backs in a pond.

Community (the **biotic community**) - all of the populations i.e. all of the living organisms within a given physical area e.g. all the animals, plants and micro-organisms within the pond.

OTHER POPULATIONS

Ecosystem - the biotic (living) community together with the physical (non-living or abiotic) environment e.g. a freshwater pond.

PHYSICAL ENVIRONMENT

A habitat must provide
- **food**
- **shelter**
- **breeding sites**

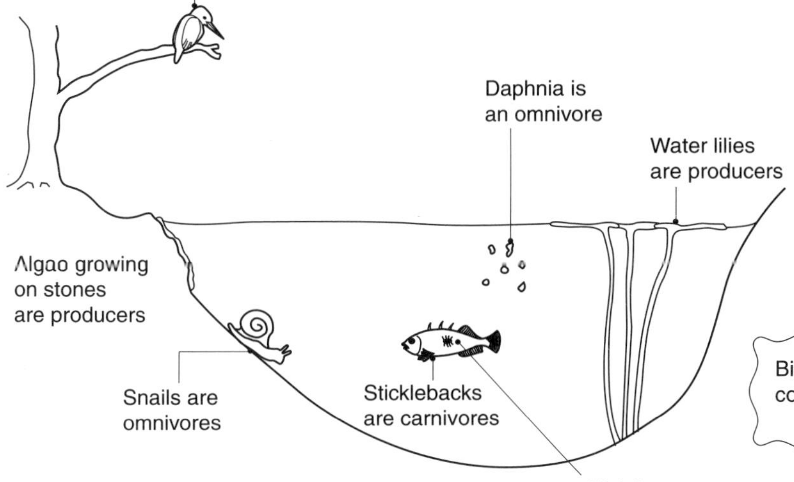

A freshwater pond ecosystem is kept stable by a series of INTERACTIONS

Biotic (living) community INTER-ACTING WITH Abiotic (non-living) factors including temperature, pH, light intensity and substrate (e.g. gravel or mud).

Autotrophs - the producers - which require an input of light and inorganic nutrients to manufacture food by photosynthesis. Include algae and water lilies.

INTERACTING WITH

Heterotrophs - the consumers - which feed either directly or indirectly on the producers. Includes herbivores, carnivores, parasites and decomposers.

Herbivores

INTERACTING WITH

Carnivores

Factors affecting population growth

Birth rate and death rate 'balance' one another resulting in an equilibrium in which the population oscillates around the carrying capacity of the environment.

FEEDBACK CONTROL OF POPULATION SIZE

'NORMAL' POPULATION → POPULATION EXCEEDS CARRYING CAPACITY → ENVIRONMENTAL RESISTANCE INCREASES → POPULATION FALLS → 'NORMAL' POPULATION

'NORMAL' POPULATION → POPULATION LESS THAN CARRYING CAPACITY → ENVIRONMENTAL RESISTANCE DECREASES → POPULATION RISES → 'NORMAL' POPULATION

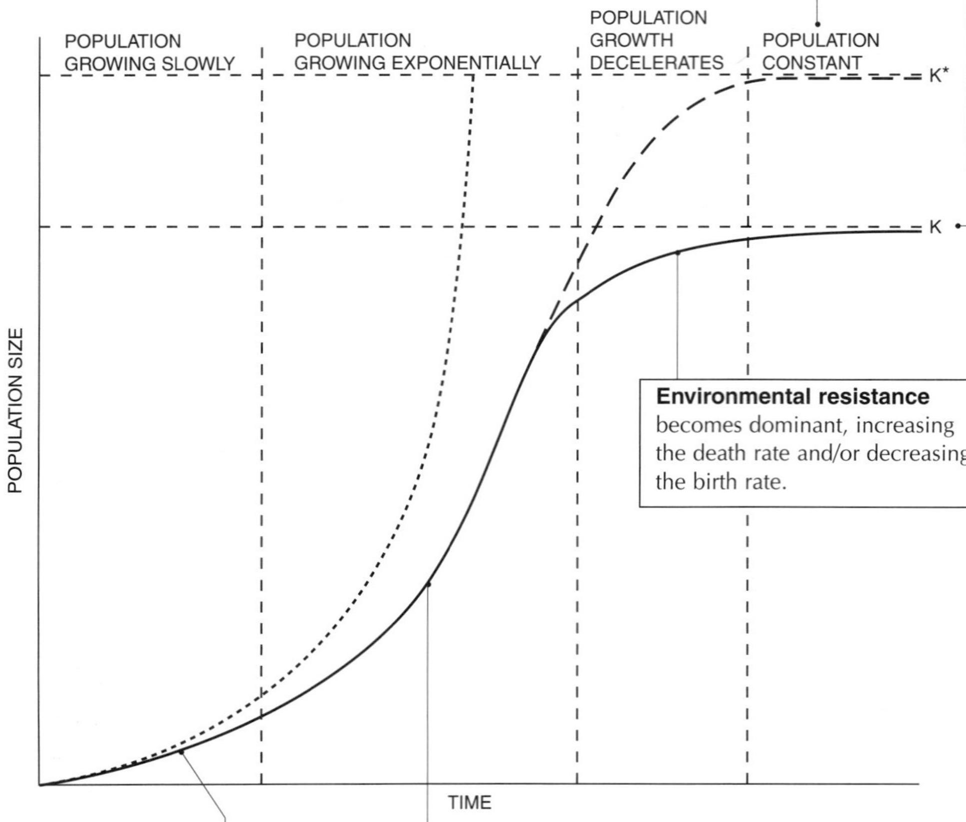

The **carrying capacity (K) of the environment** is the maximum number of a species which can be accommodated within a defined space or habitat. It is determined by the availability of **nutrients**, **shelter** and **breeding sites**. If the carrying capacity is increased (K → K*) then the size of the population tends to increase to take advantage.

Environmental resistance becomes dominant, increasing the death rate and/or decreasing the birth rate.

Human management of populations uses our understanding of these factors to

- **limit environmental resistance**
- **increase the carrying capacity**

usually for commercial gain!

Slow start to population growth as the number of mature, reproducing individuals is low and they may be widely dispersed.

Maximum observed growth rate - a compromise between biotic potential and initial environmental resistance.

- BIOTIC
  - Food availability
  - Predators or parasites
  - Disease
  - Territorial interactions
- ABIOTIC
  - Water
  - Oxygen
  - Light
  - Toxins/pollutants

Human population growth

HUMAN SUCCESS
measured as
1 worldwide distribution
2 large number of individuals
3 dominance over other species

largely due to behavioural skills which
a allow solution of complex problems
b allows control of/modification of environment

leading to **changes in carrying capacity of the environment**

THREE SIGNIFICANT PHASES IN HUMAN POPULATION GROWTH

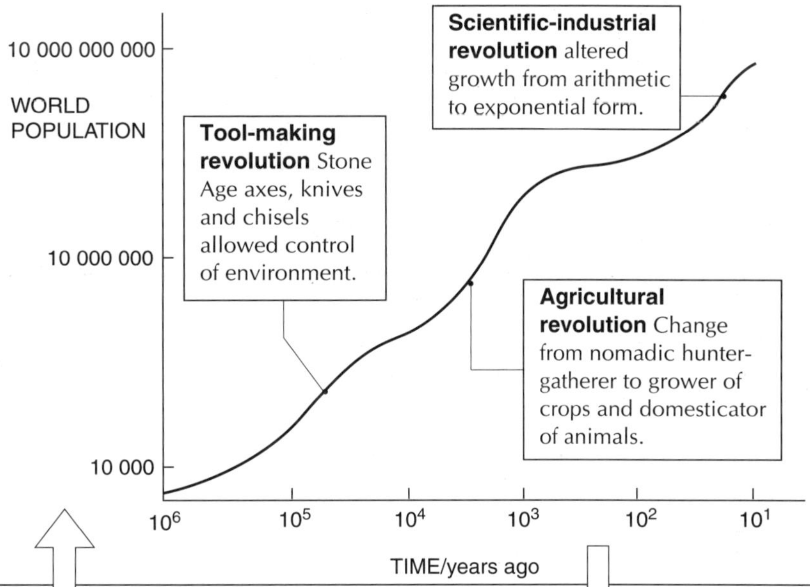

Tool-making revolution Stone Age axes, knives and chisels allowed control of environment.

Agricultural revolution Change from nomadic hunter-gatherer to grower of crops and domesticator of animals.

Scientific-industrial revolution altered growth from arithmetic to exponential form.

FACTORS AFFECTING GROWTH OF THE U.K. POPULATION

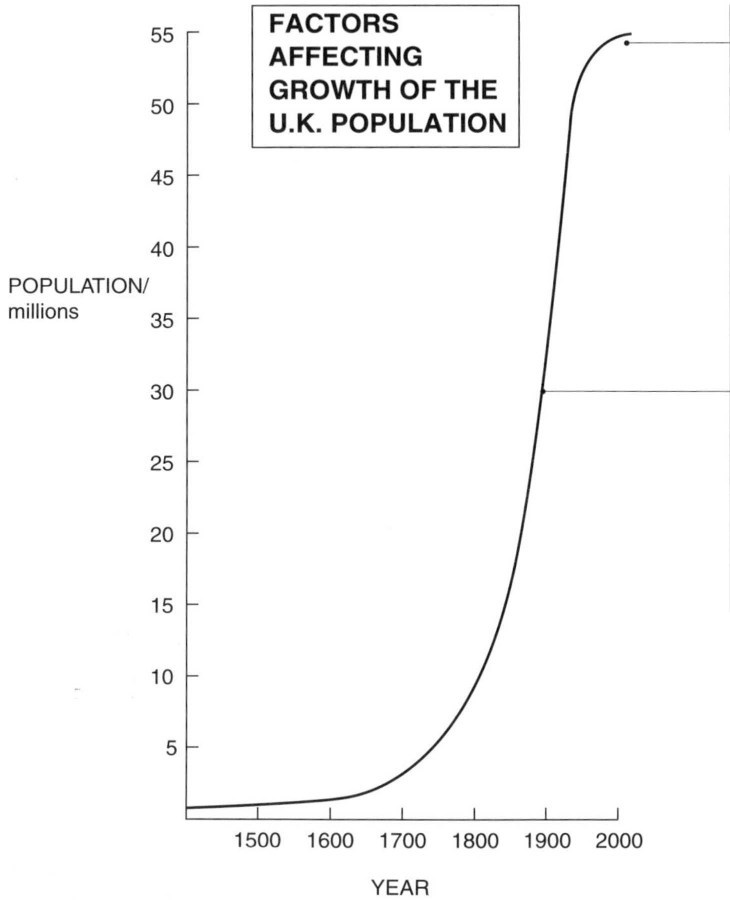

Currently static population
1 Efficient and widely-used contraception.
2 Socio-economic pressures which reduce family size.

Exponential increase in population caused by
a Increased quantity and range of food due to improved farming techniques, selective breeding of crops and animals, better storage and transport facilities and increasing imports from overseas. **Starvation no longer a limiting factor**.
b Improved hygiene reduced the incidence of water-borne diseases, and the development of vaccines and antibiotics reduced the death rate amongst individuals of pre-reproductive and reproductive ages. **Infectious disease became a less significant limiting factor**.

Pollution of the atmosphere

LEAD COMPOUNDS may slow mental development. They are in 'anti-knock' additives in petrol, and are released into the atmosphere from exhaust gases. Now most cars use lead-free petrol.

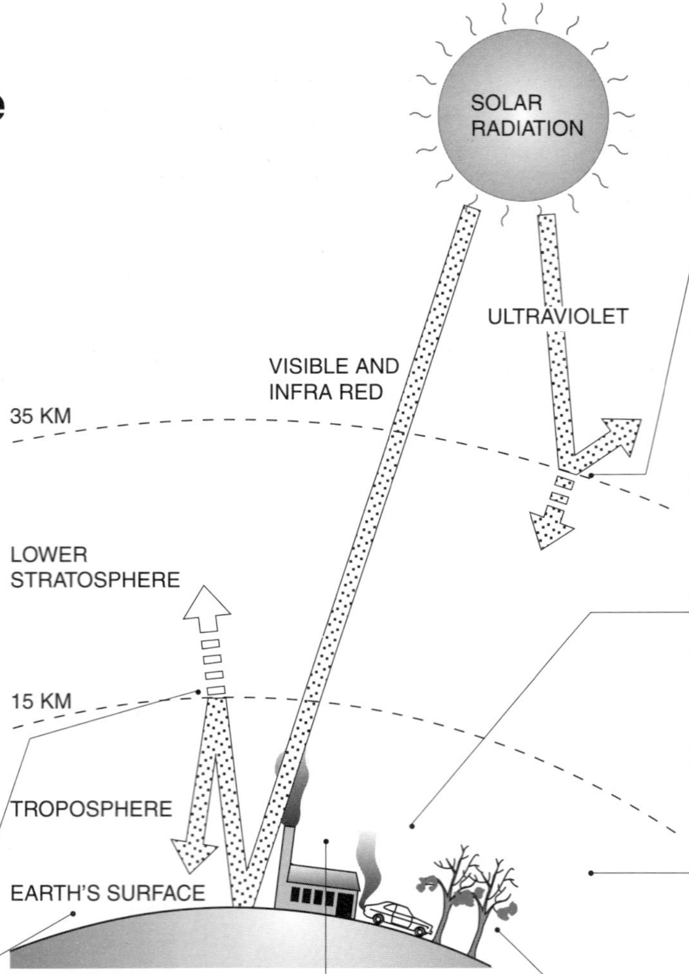

ATMOSPHERIC OZONE IS ESSENTIAL (but in the wrong place it can be harmful!)

High level ozone offers protection: ozone absorbs **solar ultraviolet radiation** which would reach the Earth's surface and cause

- sunburn
- skin cancer
- cataracts
- mutations (damage to DNA)

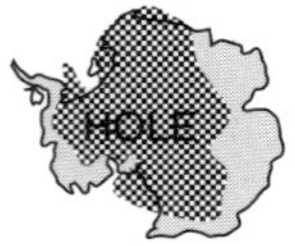

'Holes' in the ozone layer have been detected over Antarctica and are thought to have been caused by chlorine from long-lived CFCs.

Human activities affect ozone levels
CFCs (**chlorofluorocarbons**) used in aerosols, refrigerator coolants and expanded plastics **decrease amounts of high level ozone**.
Fossil fuel combustion produces oxides of nitrogen which react with oxygen **to increase amounts of low level ozone**.

Low level ozone causes problems

- acts as a **greenhouse gas**
- contributes to trapping of dust and smoke: **smog**
- causes **irritation** of eyes, throat and lungs
- damages mesophyll in leaves: **reduction in crops**.

THE GREENHOUSE EFFECT MAY CAUSE GLOBAL WARMING

Greenhouse gases include:

- **Carbon dioxide** released by combustion of fossil fuels
- **Methane** produced by ruminants - released from gut into atmosphere
- **CFC's** from aerosol propellants.

Heat cannot escape if the atmosphere contains high levels of these gases. The infra-red radiation is reflected back towards the Earth's surface.

There are good and bad results

✓ more carbon dioxide and higher temperatures mean **more photosynthesis** and **more food**.

✗ Global warming causes

- greater climatic extremes - high winds and heavier rains
- rising sea levels due to melting of polar ice
- crop losses as water evaporates from fertile areas
- extended range of pests

ACID RAIN

Human activities release acidic gases

- Sulphur and nitrogen in fossil fuels are converted to oxides during combustion.
- More oxidation occurs in the clouds, and is catalysed by ozone and unburnt hydrocarbon fuels.
- The oxides dissolve in water, and fall as **acid rain**.

$$\text{sulphur dioxide and nitrogen oxides} \xrightarrow{H_2O} \text{sulphuric and nitric acid}$$

Acid rain causes problems

- **Soils** become very acidic. This causes **leaching of minerals** and **inhibition of decay**.
- **Water in lakes and rivers** collects excess minerals. This causes **death of fish and invertebrates** so that food chains are disrupted.
- **Forest trees** suffer **starvation** because of (a) leaching of ions (b) destruction of photosynthetic tissue.

Acid rain can be reduced

- reduce emissions from car exhausts with **catalytic converters**.
- reduce emissions from power stations with **scrubbers**.
- use 'cleaner' power sources e.g. hydro- and nuclear power.

Pollution of water

PESTICIDES - overuse of pesticides on agricultural land (e.g. to protect a crop from insects) or directly on water (e.g. to kill an aquatic stage of an insect) can raise pesticide levels. The pesticide levels are then **amplified** as they pass through food chains

e.g. DDT concentration in parts per million:

WATER 0.02 — PLANKTON 5 → DAPHNIA 50 → STICKLEBACK 250 → GREBE 1500

At these concentrations the DDT is harmful and reduces breeding success.

LEAD POLLUTION - water pipes were traditionally made of lead. Lead dissolves into the water which flows through the pipes. Lead compounds are toxic and accumulate via food chains. Lead weights discarded by anglers also contribute to lead pollution.

OIL POLLUTION - oil tankers spill their contents, by accident or deliberately (during cleaning), into the sea. The oil floats on the surface, causing

- death of seabirds since feathers lose their ability to insulate when they are coated with oil.
- fish are directly poisoned
- marine mammals are killed by eating poisoned food or by loss of fur's insulating capacity.

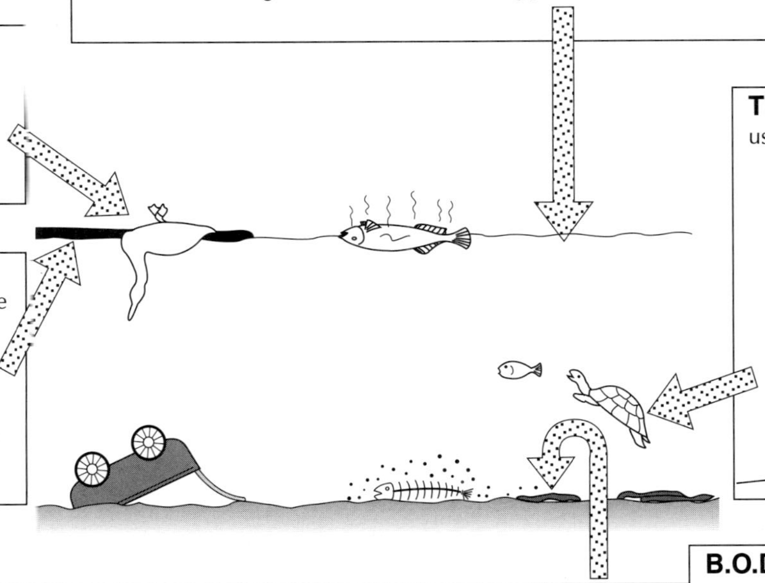

THERMAL POLLUTION - industries / power stations use water as a coolant, then discharge the water into rivers.

Now has **lower oxygen concentration** as the capacity of water for dissolved oxygen decreases as temperature rises, and as fish and bacteria become more active and so respire more. May have **new species** e.g. tropical fish, terrapins which may affect food chains.

B.O.D. is the mass of oxygen consumed by micro-organisms in a sample of water. It is determined by measuring oxygen concentration with an oxygen electrode **before and after** a period of microbial respiration. It indicates the oxygen **not available** to more advanced organisms.

EUTROPHICATION - nutrient enrichment of ponds, lakes and rivers - is responsible for BIOLOGICAL OXYGEN DEMAND

Input of raw sewage → Increased concentration of nitrate and phosphate in bodies of water

Leaching of inorganic fertilisers from farmland → Increased concentration of nitrate and phosphate in bodies of water

Increased concentration of nitrate and phosphate in bodies of water → Algae and green protists use nutrients to multiply rapidly - algal bloom

Algae and green protists use nutrients to multiply rapidly - algal bloom → Reduction of light for bottom-growing plants

Algae and green protists … → DIE → Large quantities of organic material

Reduction of light for bottom-growing plants → DIE → Large quantities of organic material

Large quantities of organic material → Aerobic decomposers (mainly bacteria) multiply and consume oxygen → Aerobic organisms (fish and invertebrates) die from lack of oxygen.

Aerobic organisms (fish and invertebrates) die from lack of oxygen. → POSITIVE FEEDBACK → Large quantities of organic material

* NB The leaching of phosphates into ponds and rivers is at least as important as nitrates in causing eutrophication

Saprotrophs cause decay

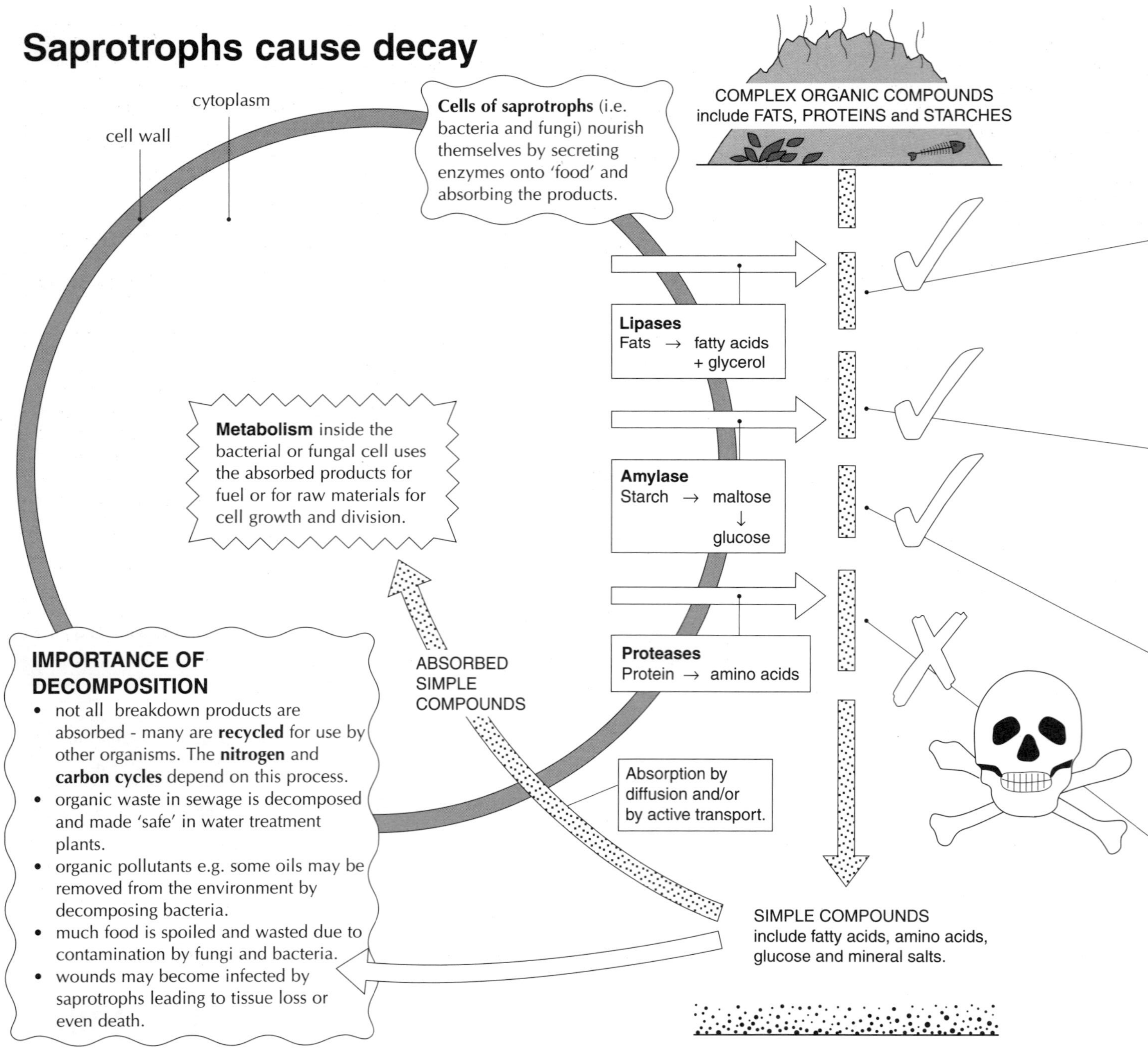

Cells of saprotrophs (i.e. bacteria and fungi) nourish themselves by secreting enzymes onto 'food' and absorbing the products.

Metabolism inside the bacterial or fungal cell uses the absorbed products for fuel or for raw materials for cell growth and division.

IMPORTANCE OF DECOMPOSITION

- not all breakdown products are absorbed - many are **recycled** for use by other organisms. The **nitrogen** and **carbon cycles** depend on this process.
- organic waste in sewage is decomposed and made 'safe' in water treatment plants.
- organic pollutants e.g. some oils may be removed from the environment by decomposing bacteria.
- much food is spoiled and wasted due to contamination by fungi and bacteria.
- wounds may become infected by saprotrophs leading to tissue loss or even death.

ENVIRONMENTAL FACTORS MAY AFFECT DECOMPOSITION

HEAT: rapid denaturation requires heat to maintain an **optimum temperature** for the activity of **enzymes**.
The heat may be generated by the respiration which occurs during the decomposition process itself.
SUGAR → CO_2 + H_2O = ENERGY (some as **heat**)

OXYGEN: is required for **aerobic respiration** which the bacteria and fungi use to release the energy needed to drive their metabolism.
Decomposition is **slow** and **very smelly** in the absence of oxygen, as methane and hydrogen sulphide may be produced.

WATER: many of the decomposition reactions are **hydrolysis reactions** i.e. use water to split chemical bonds.
Water is also necessary to dissolve the breakdown products before they can be absorbed by the saprotrophs or other organisms.

ANTISEPTICS AND DISINFECTANTS
These will kill the living organisms which carry out the decay process.

Good news! in hospitals and in for food preservation
but
Bad news! in compost heaps and in sewage works.

Treatment of sewage uses physical and biological processes to limit damage to the aquatic environment.

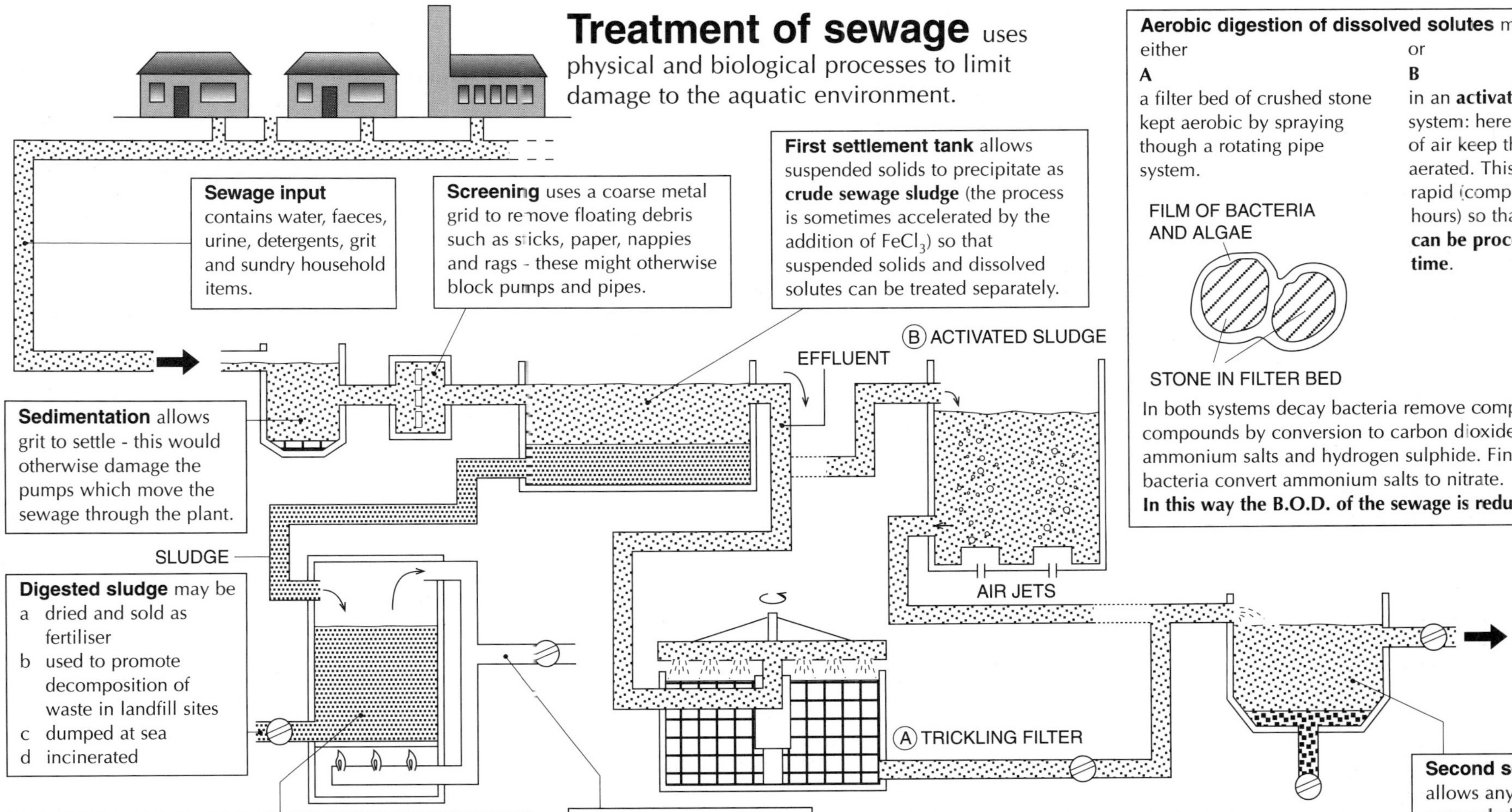

Sewage input contains water, faeces, urine, detergents, grit and sundry household items.

Screening uses a coarse metal grid to remove floating debris such as sticks, paper, nappies and rags - these might otherwise block pumps and pipes.

First settlement tank allows suspended solids to precipitate as **crude sewage sludge** (the process is sometimes accelerated by the addition of $FeCl_3$) so that suspended solids and dissolved solutes can be treated separately.

Sedimentation allows grit to settle - this would otherwise damage the pumps which move the sewage through the plant.

Digested sludge may be
- a dried and sold as fertiliser
- b used to promote decomposition of waste in landfill sites
- c dumped at sea
- d incinerated

Aerobic digestion of dissolved solutes may occur on

either
A
a filter bed of crushed stone kept aerobic by spraying though a rotating pipe system.

or
B
in an **activated sludge** system: here powerful jets of air keep the sludge aerated. This process is rapid (complete in 8-12 hours) so that **more sludge can be processed in limited time**.

In both systems decay bacteria remove complex organic compounds by conversion to carbon dioxide, water, ammonium salts and hydrogen sulphide. Finally nitrifying bacteria convert ammonium salts to nitrate.
In this way the B.O.D. of the sewage is reduced by 80-95%

Second settlement tank allows any remaining suspended solids to precipitate: the remaining effluent now has a much lower B.O.D. and a minimal pathogen count so that it can be discharged into natural waterways.

Anaerobic digestion of sewage sludge involves several stages:

1 A wide range of microbes hydrolyse

FATS	FATTY ACIDS
PROTEINS	AMINO ACIDS
CARBOHYDRATES	SUGARS

2 ... and then produce **methane**.

If these processes are not carefully regulated (they are, for example, sensitive to pH changes) they become inefficient **and very smelly**!

Methane produced during anaerobic digestion is burned to
1 power pumps and other machinery in the plant
2 raise the temperature in the anaerobic digester to 55°C - kills pathogens and speeds up the digestion process

WHY TREAT SEWAGE?

1 To remove organic compounds which might otherwise contribute to a **biological oxygen demand** of the water into which the treated sewage will eventually be discharged. The biological oxygen demand (B.O.D.) represents the oxygen used by micro-organisms and thus not available to other organisms such as fish. This oxygen consumption is very much higher if microbes multiply as they feed on organic compounds in sewage.

2 To destroy or eliminate pathogens which might be harmful to wildlife or to humans.

Ecological pyramids

Ecological pyramids represent numerical relationships between successive trophic levels in a community.

PYRAMID OF NUMBERS: a diagrammatic representation of the numbers of different organisms at each trophic level in an ecosystem **at any one time**.

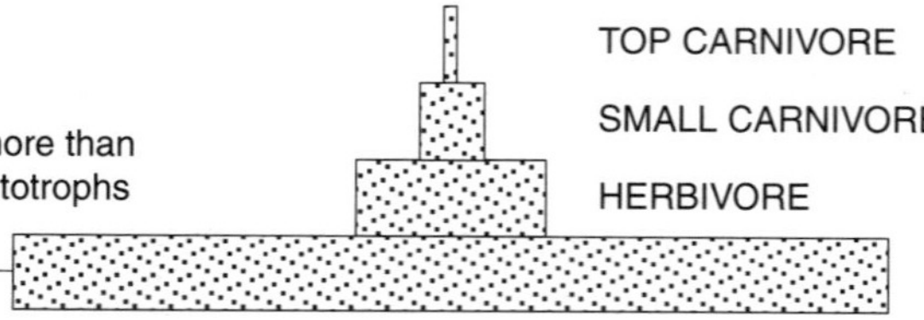

N.B.

1 The number of organisms at any trophic level is represented by the length (or the area) of a rectangle.

2 Generally, as the pyramid is ascended, the **number** of organisms **decreases**, but the **size** of each individual **increases**.

BUT WAIT!

PROBLEMS

a The range of numbers may be enormous - 500 000 grass plants may only support a single top carnivore - so that drawing the pyramid may be very difficult.

b Pyramids may be **inverted**, particularly if the **producer is very large** (e.g. an oak tree) or **parasites feed on the consumers** (e.g. bird lice on an owl).

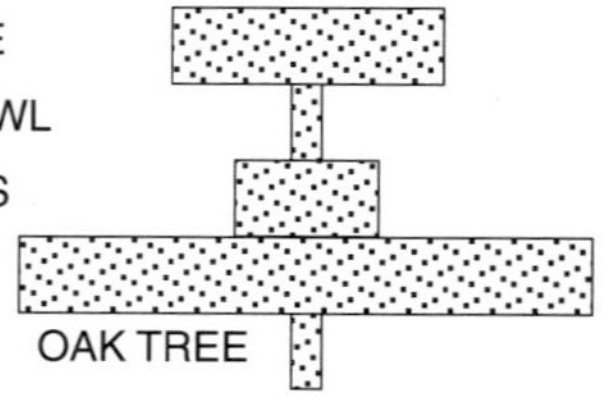

SO...

PYRAMID OF BIOMASS...

...which represents the **biomass** (number of individuals x mass of each individual) at each trophic level **at any one time**. This **should** eliminate the inversion and scale problems encountered when constructing a pyramid of numbers.

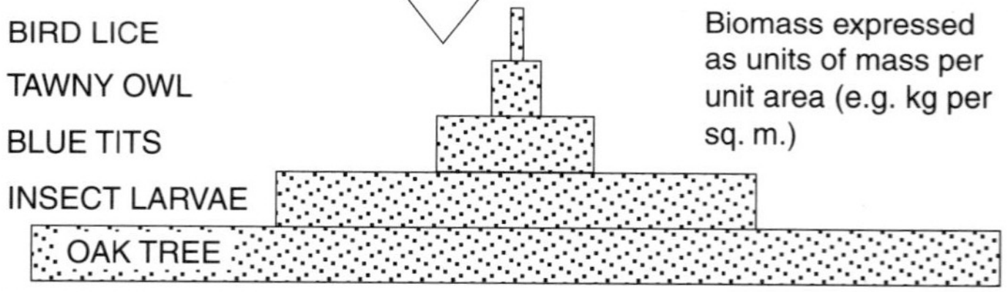

FOOD CHAINS represent a **flow of energy** between successive **trophic levels**.

Primary production: plants are able to 'fix' about 5% of the light energy falling on them into chemical energy in food. Much light is reflected, or is not of the correct wavelength for photosynthesis.

Energy transfer to primary consumer is between 5 and 10%
- much of plant body is indigestible e.g. cellulose and lignin.
- consumer rarely eats whole plant - roots and stems may be left behind.

Energy transfer to secondary consumer is between 10 and 20%
- animal material has a higher energy value
- animal material is more digestible

Respiration losses occur from each trophic level. Respiration releases the energy necessary to drive metabolic reactions e.g. active transport and protein synthesis.

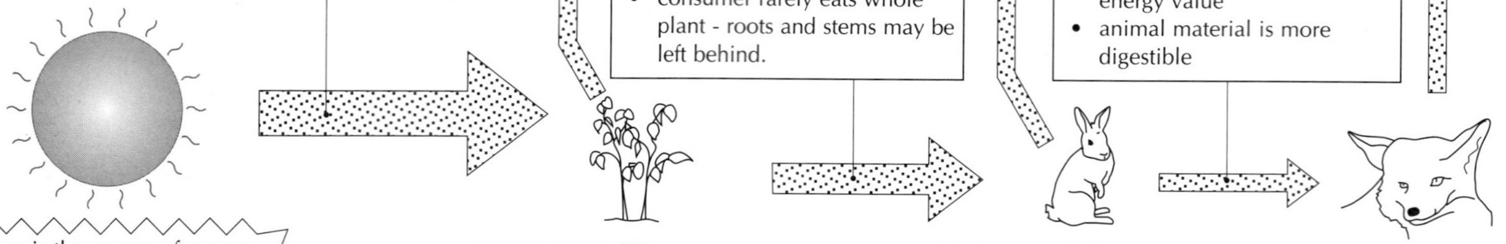

The Sun is the source of energy which drives all food chains. Less than 1% of the energy released from the Sun's surface actually falls onto the leaves of the photosynthesising plants.

Decomposer organisms are fungi and bacteria which obtain their energy and raw materials from plant and animal remains - eventually the entire energy content of these remains will be released as **heat** from inefficient respiration. Some energy may be 'stored' in fossil fuels e.g. coal, if decomposition cannot be completed.

Food chains are short because
- long chains would need an enormous producer biomass spread over a vast area
- energy transfer between levels is so inefficient - fox might only receive 10% x 10% x 5% of light energy falling on producer.

Biological pest control

depends on an understanding of the relationship between populations of predators and their prey.

IDEAL RELATIONSHIP BETWEEN PEST AND ITS CONTROL AGENT

The **pest species** becomes the **prey** of the control agent: it is the **target** in the system of biological control.

Control agent is a natural **predator** on the **prey** pest species - population increases as the predator breeds.

Introduction of control agent: must be a population large enough to control the pest before it causes too much damage.

Pest population falls due to **predation** by control agent.

Population size above which the pest is **economically harmful**: often determined by the expected yield and potential value of the crop.

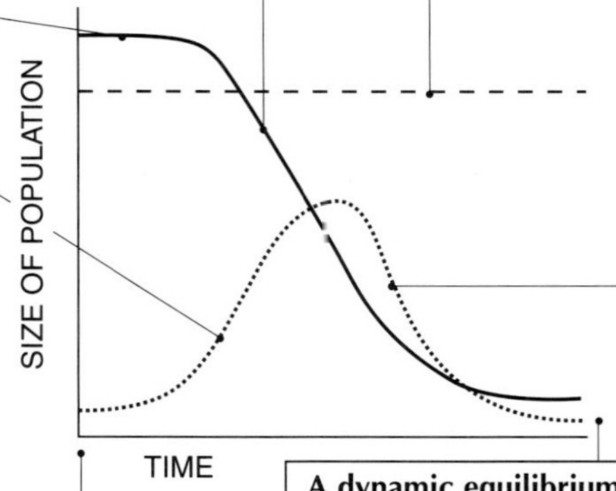

Population of control agent falls because of food shortage caused by a reduction in prey (pest) numbers.

A dynamic equilibrium is set up in which a moderate residual population of the control agent is able to permanently restrict the population of the pest. NB the pest species **must not be entirely eliminated** or the control agent will die out and a further introduction will be necessary to prevent re-establishment of economically damaging pest populations.

PRINCIPLE TECHNIQUES IN BIOLOGICAL CONTROL

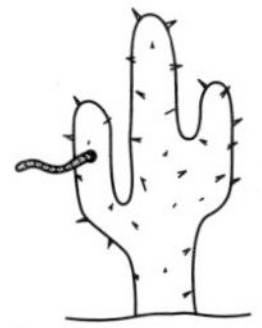

1. Use a **herbivore** to control a **weed** species e.g. *Cactoblastis* larvae on prickly pear.
2. Use a **carnivore** to control an **herbivorous pest** e.g. hoverfly larvae on aphids.
3. Use a **parasite** to control its **host** e.g. *Encharsia*, a parasitic wasp, on the greenhouse whitefly, *Trialeurodes vaporariorum*.
4. Disrupt the **breeding cycle** of a pest **if it mates once only in its life** e.g. release of irradiation-sterilised males of the screw worm fly, a flesh eating parasite of cattle.

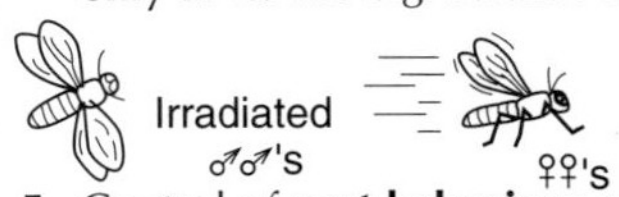

5. Control of **pest behaviour** e.g. sex attractant pheromones are used to attract apple codling moths into lethal traps.

... BUT THERE HAVE BEEN FAILURES!

Hawaiian cane toads introduced into Queensland, Australia, to control the greyback beetle, a pest on sugar cane, are now a serious threat to Australian wildlife. The toad

- eats many native insects and worms
- displaces native frogs and toads from breeding pools
- poisons larger animals which try to eat it because its skin is extremely toxic.

Cats and stoats introduced to offshore islands of Britain and New Zealand to limit population of rodents which threatened rare ground-nesting birds found that the chicks of the 'protected' birds were easier to catch than the rats were!

NEVERTHELESS, COMPARED WITH CHEMICAL CONTROL, THERE ARE CERTAIN ADVANTAGES OF BIOLOGICAL PEST CONTROL...

	BIOLOGICAL CONTROL	CHEMICAL CONTROL
SPECIFICITY	Very high, so little danger to beneficial species or to humans	May be low, but legislation leading to great improvements
'ACCUMULATION' IN ECOSYSTEMS	None	Concentrations may increase along food chains
PERMANENCE OF CONTROL	Good, but small numbers of pests must be tolerated	Requires regular reapplication
DEVELOPMENT OF PEST RESISTANCE	Very rare	Common, requiring ever-increasing 'dose'
COST IN FINANCIAL TERMS	Initially may be high, but very low in the long-term	May be very high, restricting use to wealthy nations

Cycling of nutrients involves **interconversion of simple and complex molecules**

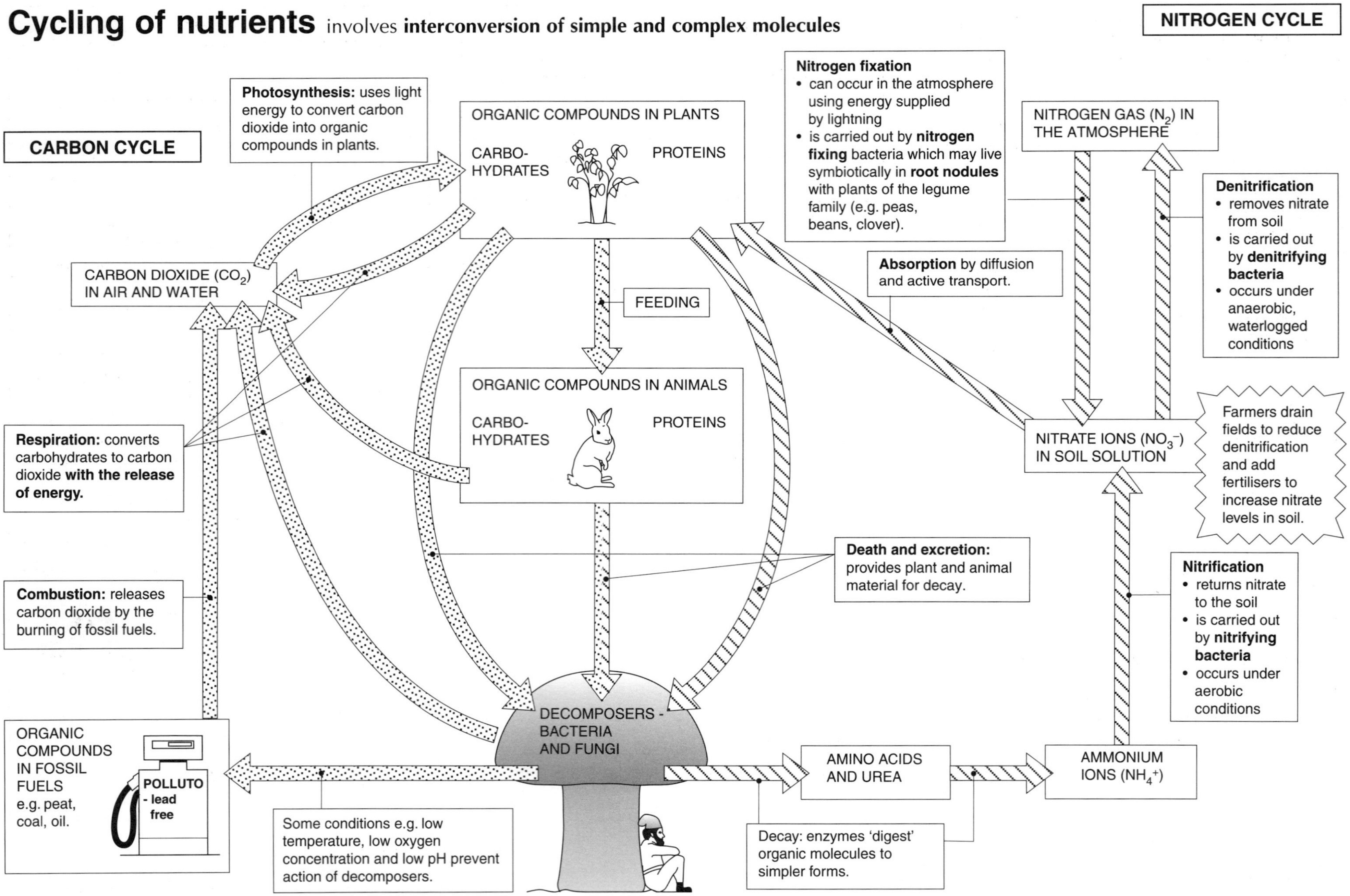

Managing ecosystems: fish farming

maximises profit by minimising environmental resistance

AGE FOR HARVESTING

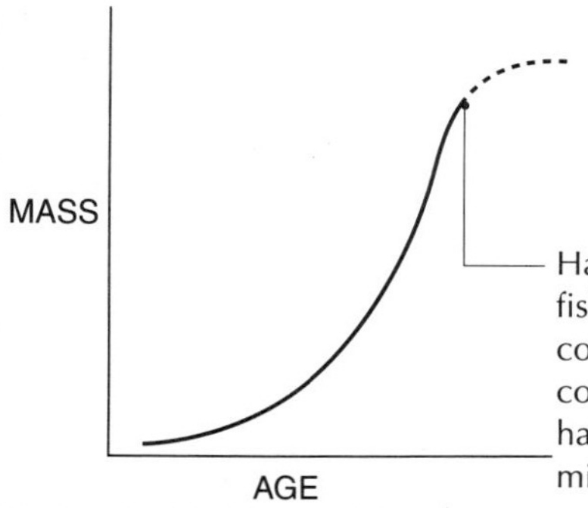

SUSPENDED NET

- keeps out **aerial predators** such as cormorant, heron and osprey.

PREPARED FOOD

- dried, in pellet form, for convenience of transport from site of production and for measured delivery to fish.
- usually made from 'trash fish' which is marine fish caught in nets but not used for human consumption – **very high in protein** (rapid growth rate) but **very expensive**.
- may include **colouring agent** (turns fish pink – favoured by consumers) and **antibiotic** (disease control).

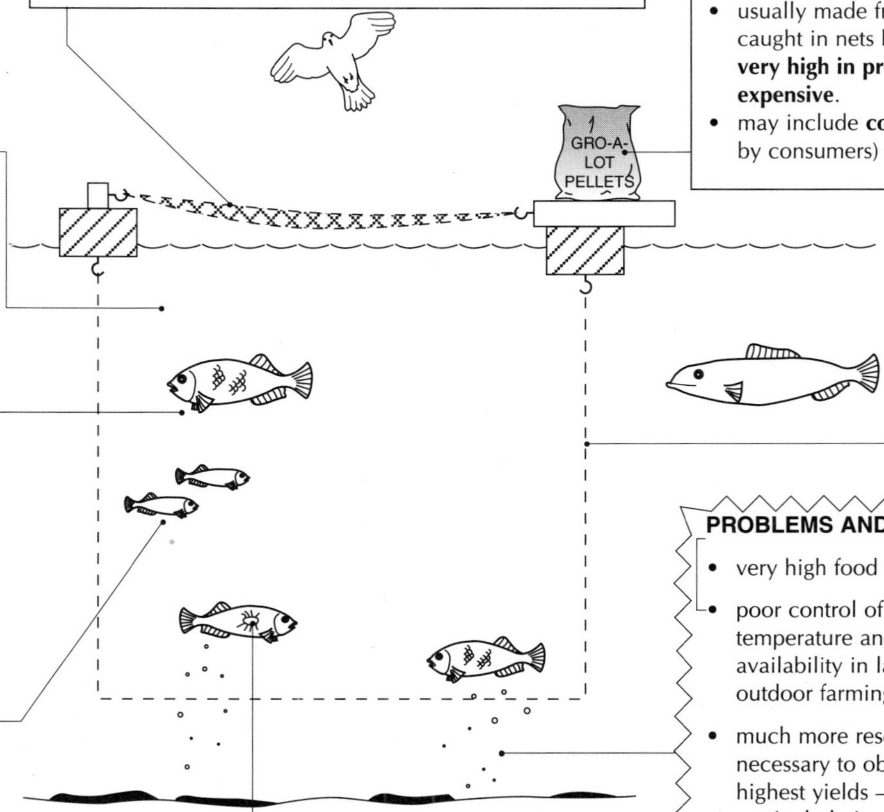

SPECIES – the 'farmed' species must

- grow well under captive conditions
- accept prepared i.e. non-living food
- have a high **conversion ratio** i.e. convert food → flesh efficiently
- ideally be able to complete its life cycle i.e. breed under captive conditions
- not be particularly susceptible to disease.

It is usually **expensive** to provide these conditions so the product should **command a high price**. For this reason the main farmed species **in Britain** are members of the Salmon family.

HANGING NET

- keeps out **aquatic predators** such as otter and pike
- keeps out other fish and so **reduces competition** for the pelleted food.

FRY PRODUCTION

- fish are spawned under artificial conditions - often in aquaria and often under the influence of reproductive hormones.
- fry are 'grown on' to a size at which they can be released
  - control temperature (higher temperature, rapid growth)
  - high oxygen levels with aerators
  - growth hormones may be added to water
- fry of **uniform size** are released into the farming pens. This reduces the chance of the fish eating each other!

DISEASE CONTROL: close confinement makes fish more likely to suffer from disease ('spread' is much easier)

- dose water with **dichlorvos** which kills **fish lice.**
- dose water with **fungicide** to prevent **fungal** infection of skin and gills.
- add **antibiotic** to food to control **bacterial** infections.

PROBLEMS AND ENVIRONMENTAL CONCERNS

- very high food costs.
- poor control of temperature and oxygen availability in large outdoor farming pens.
- much more research necessary to obtain highest yields – particularly important are **selective breeding** programmes to develop new fish varieties with improved growth rates and conversion ratios.

Δ pollution by pesticides since these compounds may kill organisms which are foods for 'wild' species.

Δ excess food and fish faeces create nutrient rich environment below netted area → growth of bacterial population → increased **biological oxygen demand**.

In developed countries intensive fish farming is so costly that it should be seen as **adding variety to the diet** rather than being a **source of cheap, plentiful protein**.

Managing ecosystems: horticulture

can be profitable if environmental resistance is reduced

CARBON DIOXIDE CONCENTRATION: is a major limiting factor in photosynthesis. In a greenhouse the plants may photosynthesise very quickly and CO_2 is rapidly used up. The CO_2 concentration is usually raised to about 0.1% of the atmosphere (about three times that in air). This gives a significant (about 50%) increase in crop yield. The extra CO_2 can be provided by burning paraffin (which also raises the temperature) or, more accurately, by releasing it from a cylinder.

COMPUTER CONTROL: this is widely used in large commercial greenhouses. Sensors provide information about air temperature, CO_2 concentration, water available to roots, humidity and mineral concentration, and the control centre ensures that any changes are corrected.

e.g.

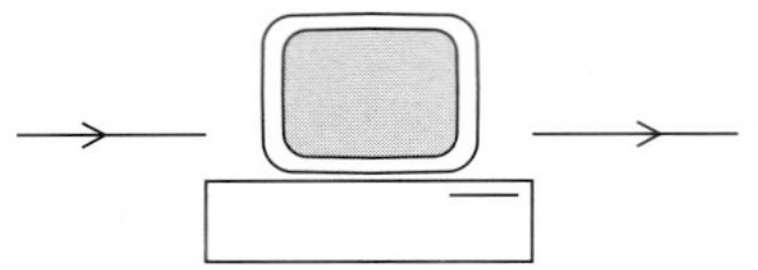

HIGH-YIELDING STRAINS OF CROP: selective breeding and/or genetic engineering can develop crop strains which

- have a high yield
- produce fruit of a desirable colour/texture
- produce fruit at the same time
- may have genetic pest resistance.

POLLINATING INSECTS: bees may be introduced to ensure high rates of fertilisation and therefore fruit formation.

HUMIDITY: this affects the rate of transpiration, and therefore the rate at which the plants must be supplied with water. Thus a high humidity reduces the need for additional water and generally favours growth (NB **too** high humidity favours the growth of fungal pests). Humidity is reduced by opening ventilators and raised using an automatic mist spray.

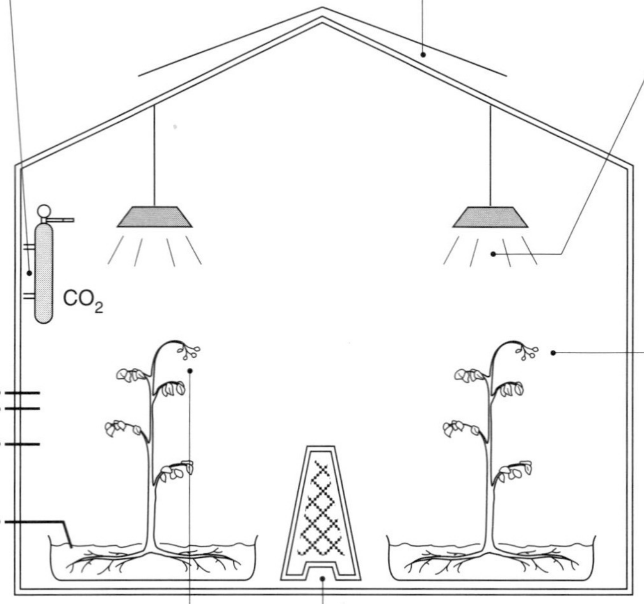

TEMPERATURE: affects plant growth because of its affect on the enzymes of photosynthesis. NB high temperatures may also speed up the life cycle of pests. Sunlight provides some heat (shading may be necessary in summer) but more control is available using thermostatically-regulated heaters.

ILLUMINATION: it is important to control

- **intensity** - this will influence the rate of food production by photosynthesis. Greater intensity → more photosynthesis until some other limiting factor intervenes.
- **quality** - photosynthesis is most efficient at red and blue wavelengths so that 'white' light contains some wavelengths ('green') which are not useful.
- **duration** - if fruit is the desired product the plant must flower. Flowering is controlled by daylength (the duration of light in a 24 hour period).

Sunlight provides some illumination but artificial lighting systems are more controllable (but more expensive).

PEST CONTROL: is particularly important since a pest species could spread very rapidly through a greenhouse containing only a single crop.

- **Fungi** can be a problem since high humidity/high temperature encourages spore germination and growth of fungus. Control with **fungicide spray** or by **reduction of humidity**.
- **Weed plants** could compete with the crop species for light, water and mineral nutrients. They are removed **by hand** or with a **selective herbicide**.
- **Animal pests** are herbivorous (e.g. caterpillars) or sap-sucking (e.g. aphids) insects. These may be controlled using **insecticides** but
  - these can be inefficient, especially against waxy-coated pupae
  - they may need to be reapplied, so can be expensive
  - they may leave residues so that crop will need washing

  or **biological control** which
  - uses a natural predator of the pest
  - usually only requires a single application
  - does not leave residues

  Important examples of biological control are
  ladybirds which eat aphids
  wasp larvae which live as parasites on **whitefly larvae**

Managing ecosystems: animal husbandry.

Intensive farming aims to minimise cost/maximise productivity by reducing environmental resistance.

Temperature control: essential so that costly heat energy is not wasted.
if **too high**, animals are uncomfortable and will not feed
if **too low**, food intake is 'wasted' on heat production to maintain body temperature.

Controlled light regime/photoperiod: may influence growth rate since can permit longer feeding period
may control reproductive cycle so, for example, milk production may be stabilised through the year.

Shelter:
Prevent entry of **predators**
Eliminate/control **competitors** for food
Protect against **climatic extremes**

Food input:
Control content
- high protein for growth
- minimal fat to suit customer demand for lean meat
- include growth hormone to increase growth rate
- add copper ions which reduce energy consumption for heat production

Often use dried milk powder or single cell protein, with mineral and vitamin supplements.

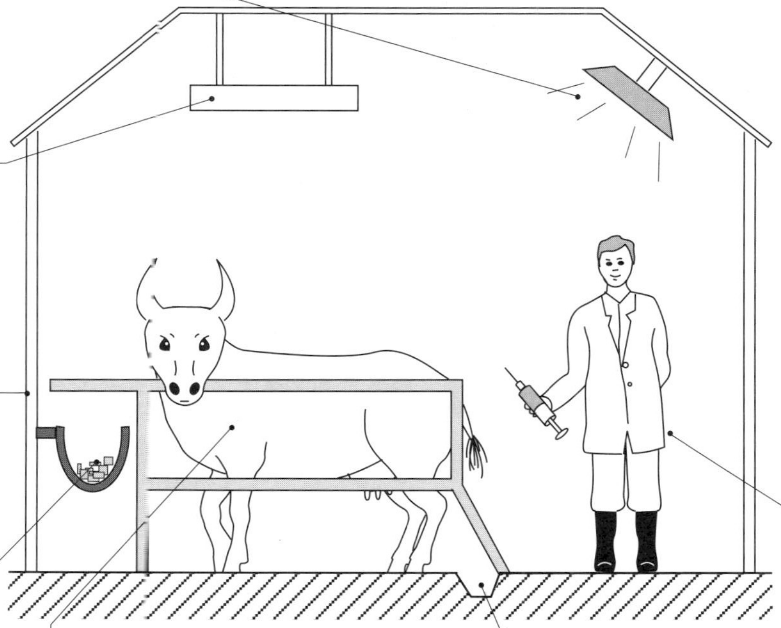

Strain of animal: selective breeding for animals with high **conversion ratio** i.e. most efficient transfer of food intake to body mass.

Minimise movement: less energy consumption and thus more efficient 'conversion'.
NB can upset **social interactions**, and stress can lead to poor growth/unpleasant 'cortisone' taste.

Hygienic conditions: most animals are free of gut parasites and debilitating bacteria - healthy animals grow more quickly and meat is more saleable.

Slurry (faeces/urine) can be dried and recycled for use as fertiliser.

GROWTH CURVE may indicate optimum time for marketing animals

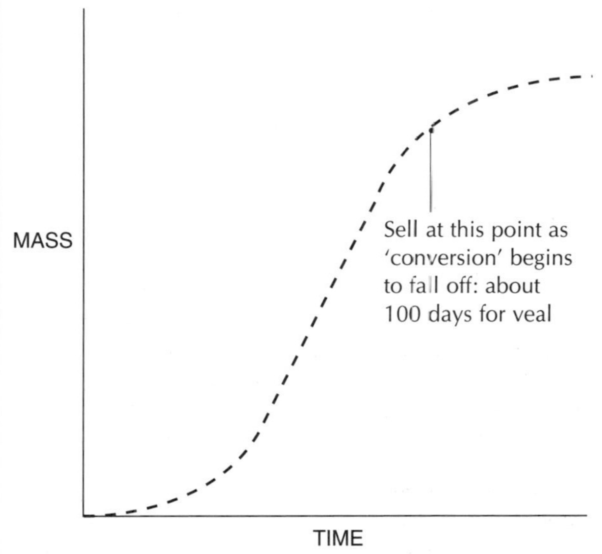

Veterinary care:
Antibiotics to reduce bacterial infections
Vaccination to minimise viral infections
Hormone/vitamin supplements can be administered more accurately than in the diet
Artificial insemination techniques can reduce costs (no need to keep bulls in dairy farms).

Design an experiment

An experiment is designed to **test the validity of an hypothesis** and involves the **collection of data**

e.g. light intensity affects the rate of photosynthesis

using appropriate apparatus and instruments

The volume of oxygen released in a fixed length of time can be used to calculate the rate of photosynthesis: this is the **dependent (responding) variable**.

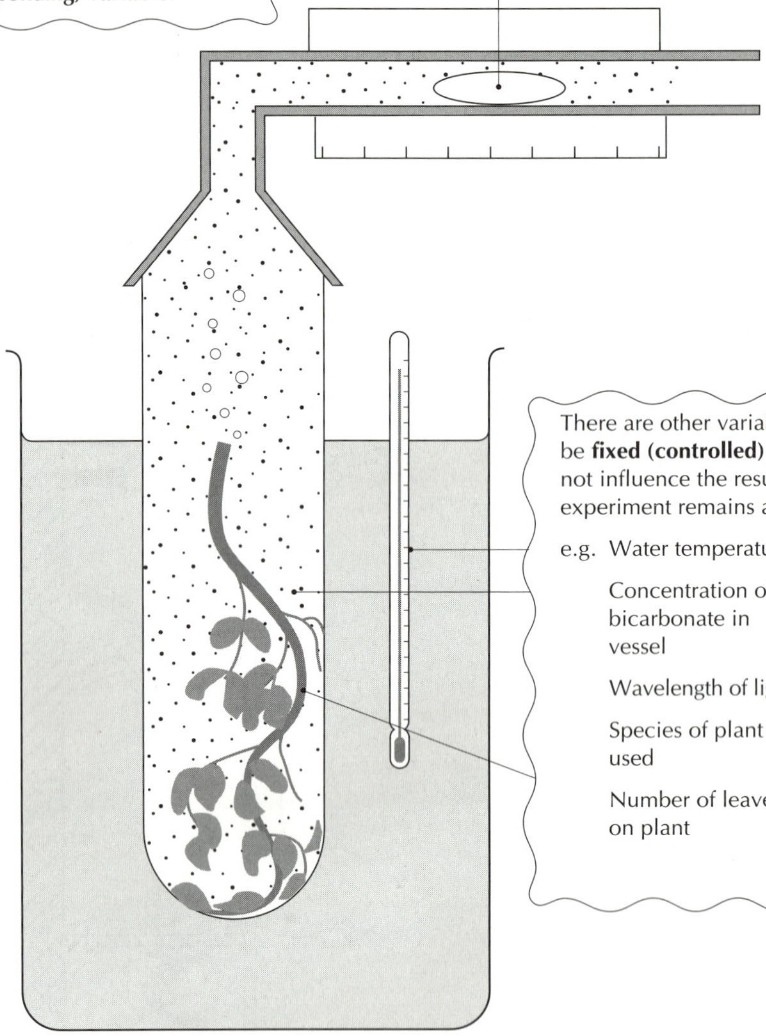

The light intensity can be varied by the experimenter: this is the **independent (manipulated) variable**.

Light intensity is an example of a **continuous variable**.

There are other variables which must be **fixed (controlled)** so that they do not influence the results, and the experiment remains a **fair test**

e.g.	Water temperature	(continuous)
	Concentration of bicarbonate in vessel	(continuous)
	Wavelength of light	(continuous)
	Species of plant used	(categoric)
	Number of leaves on plant	(discrete)

A **control** experiment is the same in every respect **except** the manipulated variable is not changed but is kept constant.

A control allows confirmation that **no unknown variable** is responsible for any observed changes in the responding variable: it helps to make the experiment a **fair test**.

A **'repeat'** is performed when the experimenter suspects that misleading data has been obtained through "operator error".

'Means' of a series of results minimise the influence of any single result, and therefore reduce the effect of any 'rogue' or anomalous data.

Dealing with data

may involve a number of steps

1 Organisation of the **raw data** (the information which you actually collect during your investigation).

2 Manipulation of the data (converting your measurements into another form).

and

3 Representation of the data in **graphical** or other form.

Steps 1 and 2 usually involve **preparation of a table of results**

Raw data: independent variable always in LH column

May often involve complex units: take care in presentation e.g. beats per minute or beats/min

Column heading should include **name of variable** and **units** e.g. **temperature/°C** or **temperature in °C**.

INDEPENDENT or MANIPULATED VARIABLE e.g. TEMP/°C	DEPENDENT or RESPONDING VARIABLE e.g OXYGEN GIVEN OUT/cm^3	MANIPULATED DATA e.g. RATE OF OXYGEN PRODUCTION/cm^3 per min.
20.00		
26.25		
–		
30.75		

Derived from the raw data e.g. if 3 readings of a responding variable are taken, the **mean** of these readings would be manipulated data OR if you measured heart rates in beats per 15 seconds, the rate in beats per minute would be manipulated data

Units should **always** be in column heading and **never** alongside the numerical value.

Present **numerical data**

a in decimal form e.g. 6.25 not $6\frac{1}{4}$

b to the same number of decimal places e.g. if 6.25 given then use 6.00 not 6

c as true numbers e.g. 0.7 not .7

There should be an **informative title:** simple **effect of** (variable in LH column) **on** (variable in RH column)
e.g. Effect on temperature on pulse rate in humans

There should be no 'blanks':

a **missing value** should be shown as –

b **value of zero** should be shown as 0

Rule lines around table: looks **neater** and **more professional**.

Graphical representation: A graph is a visual presentation of data and may help to make the relationship between variables more obvious.

For example

AIR TEMPERATURE /°C	BODY TEMPERATURE OF REPTILE /°C
20	19.4
25	25.9
30	30.4
35	35.1
40	40.1
45	44.8

isn't as helpful as

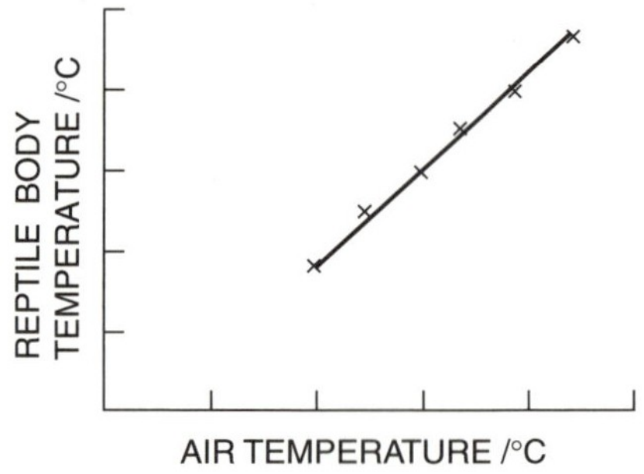

which suggests that **reptile body temperature is directly proportional to air temperature**

A graph may be produced from a table of data **following certain rules**

y axis

dependent (responding) variable

NB units separated from physical quantity by a solidus

e.g. **reptile temperature /°C**

or in words

e.g. **reptile temperature in °C**

There should be an **informative title:** simply **Effect of** (variable on *x* axis) **on** (variable on y axis).

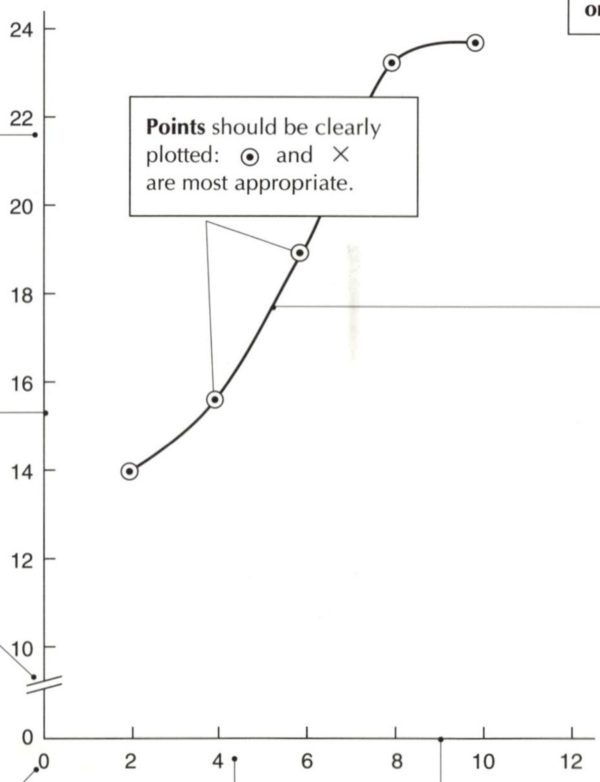

Points should be clearly plotted: ⊙ and × are most appropriate.

The line of the axes should be ruled in: use a **black** line.

Axes should be **linear** and, wherever possible, **unbroken**. If an axis must be broken to make best use of the scale, the break should be shown as ⫲

Origin should be shown on both axes.

Scale on axes should
a be marked **in equal increments**
b extend to cover all plotted points
c make best use of the space available

x axis
independent (manipulated) variable
physical quantity/units

Once the points are plotted they may be joined by a smooth curve, if theory predicts a smooth, gradual transition…

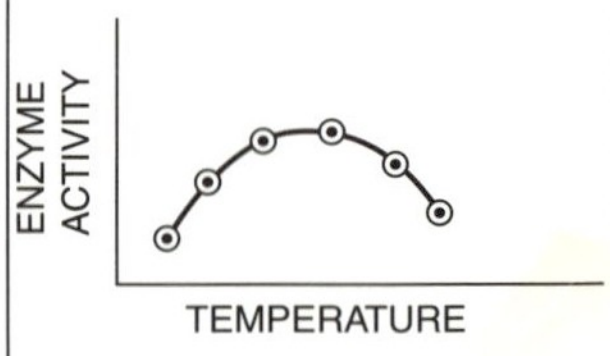

… or by **short, straight lines** if a smooth, gradual change cannot be predicted.

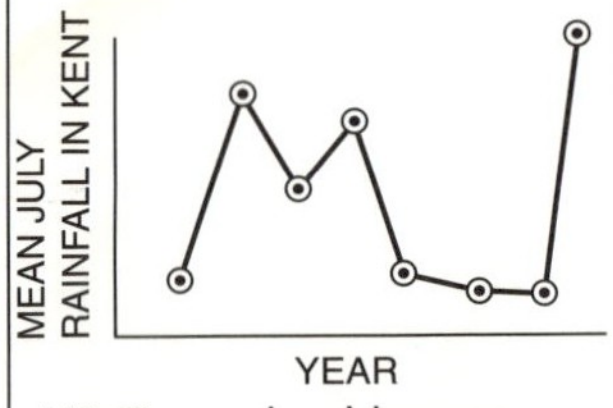

NB Curve should **never** extend beyond final plotted point.

Sample exam question answers

On the following pages some sample examination questions are answered. As you study them try to notice the following key points:

- what are you told to do and has the answer done it? key words to look for are: ***list; describe; explain; suggest; predict***
- the number of marks offered must be matched by the number of points made
- can you recognise what part of the syllabus is covered by the question because this will guide you to the likely answer
- some questions cover complicated material but actually contain most of the information which you need to gain full marks!

7 a Restriction endonucleases are enzymes used to cut DNA into fragments during genetic engineering. EcoR1 is a restriction enzyme produced by the bacterium *Escherichia coli*. It always cuts DNA at the same sequence of organic bases. The base sequence below shows some of the remaining organic bases after a cut.

Write the missing bases in the correct position to show the DNA before the cut.

many candidates missed this instruction (and hence 1 mark)

G A A T T C

C T T A A G (1)

*could be easily worked out as long as you noted the question 'gave' the pairing of G-C, the other base pair therefore **must** be A-T (or T-A)*

b The diagram below represents a length of DNA after being treated with EcoR1 restriction endonuclease.

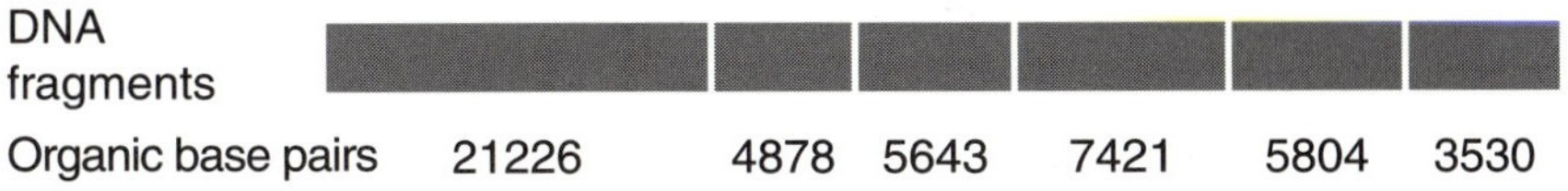

(i) How many C T T A A G base sequences were on the original DNA molecule?

5 (1)

ie. there were five points which this enzyme could recognise and 'cut' the DNA

(ii) Work out how many individual organic bases there are on the DNA shown.

48502 x 2 (1)

*question asks for **individual** bases, there are two in each pair!*

*total number of **base** pairs (21226 + 4878 + 5643 + 7421 + 5804 + 3530)*

(c) The following table lists events from the identification of a human gene coding for a hormone "X" to the commercial production of hormone "X".

Genes can be transferred into plasmids, tiny circles of DNA which are found in bacteria.

Show the correct sequence of events 1 to 8 by writing the appropriate number in each box provided. The first (number 1) and last (number 8) have been completed for you.

follow very straightforwardly from number 1.

Event	Number
Cutting of a bacterial plasmid using restriction endonuclease	3
Cutting of human DNA with restriction endonuclease	2
Identification of the human DNA which codes for hormone "X"	1
Many identical plasmids, complete with human gene, are produced inside the bacterium	6
Mixing together human gene and "cut" plasmids to splice the human gene into the plasmid	4
Some of the cloned bacteria are put into an industrial fermenter where they breed and secrete the hormone	8
The bacterium is cloned	7
Using the plasmid as a vector, inserting it, complete with human gene, into a bacterium	5

clearly the step prior to number 8

Six marks gained by applying a logical and methodical approach to dealing with information supplied by the examiner.

(6)

(Edexcel)

1 The table below shows how three alleles control the ABO human blood group system.

Allele	Effect
I^A	Causes the production of antigen A on red blood cells
I^B	Causes the production of antigen B on red blood cells
I^O	Does not cause the production of antigens on red blood cells

Important information which will be used later in the question

When I^A and I^B are inherited together they are equal in their effect. When I^O is inherited with either I^A or I^B it is recessive.

A clear instruction, no call for explanation here

(a) Complete the diagram below by filling in the blank circles to show the possible inheritance of the ABO blood group system in Mr and Mrs Smith's children.

At meiosis (reduction division) only one of each pair of alleles can be passed to any single gamete

At fertilization two gametes fuse to form a zygote

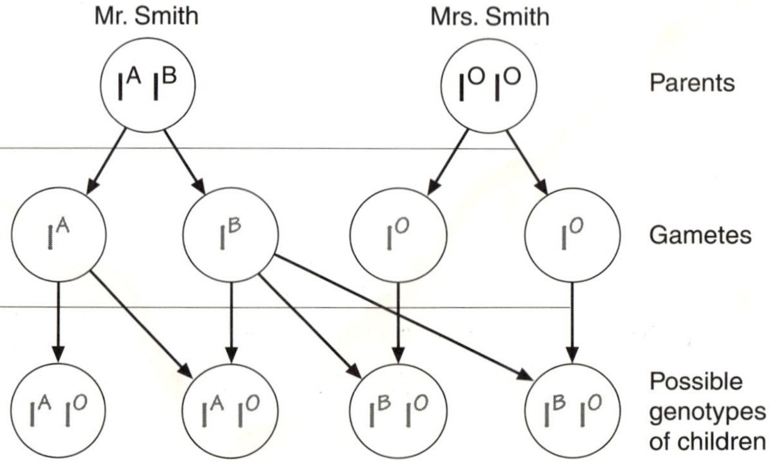

(b) When I^A and I^B are inherited together they both show their effects in the phenotype. Name the term used to describe this.

Co-dominance (1)

Important! no call for any explanation

(c) **(i)** What is the probability of the eldest child being male?

50% or $\frac{1}{2}$ or 1 in 2 (1)

Sex is determined by X or Y chromosome carried in the male gamete, there are equal numbers of 'X' and 'Y' gametes.

(ii) What is the probability of the eldest child being male with blood group B?

$\frac{1}{2}$ (being male) x $\frac{1}{2}$ (being B) = $\frac{1}{4}$ or 25% or 1 in 4 (1)

(N.B. **not** 1:4, this is 1 in 5 or 20%)

(d) Sometimes a patient needs a blood transfusion. It is very important that the blood given to the patient does not contain any antigen which is not in the patient's own red blood cells. Which parent can donate blood safely to any of the possible children?

Mrs. Smith (Since I^O blood has no damaging antigens) (1)

(Total 7 marks)

(Edexcel)

2 The diagram below shows a section through a leaf.

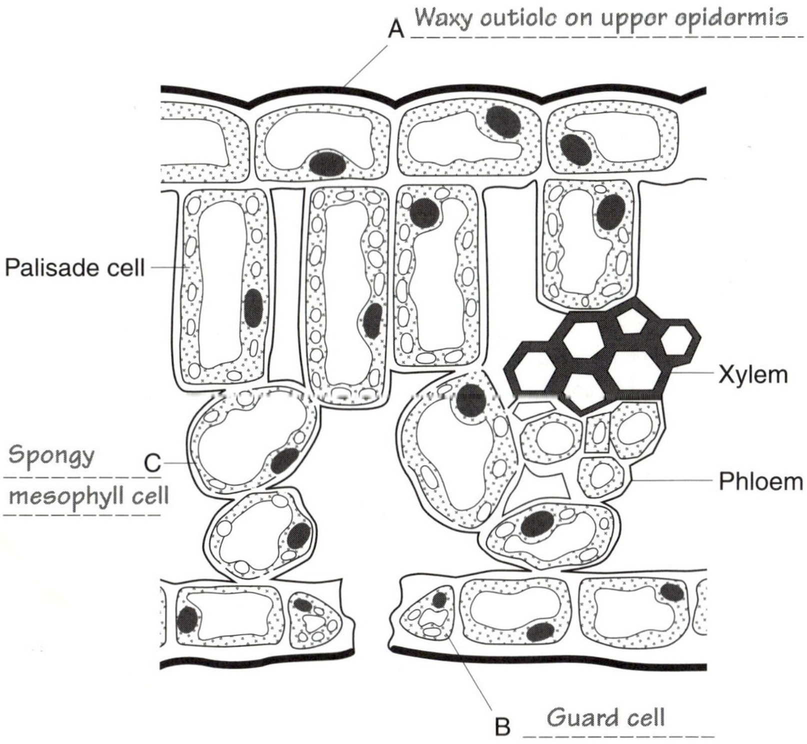

(a) Label layer A and cells B and C on the lines provided. (3)

(b) Which cells of the leaf contain chloroplasts?

Palisade cell: spongy mesophyll cell: guard cell (1)

notice the 'plural' ie more than one type of cell

you can see the chloroplasts on the diagram!

no call for explanation or comment

(c) Describe how water reaches a leaf and enters a palisade cell.

Evaporation/transpiration from leaf surface
water pulled up xylem to replace 'losses'
water moves by osmosis to palisade cell
water crosses palisade cell membrane by osmosis (3)

make **three** points for the **three** marks on offer

(d) Describe how sugar produced in a palisade cell reaches the roots.

Conversion to sucrose/transported in phloem/
unloaded in roots by diffusion or active transport (2)

more factual recall, **but must not be glucose!**

(e) The sugar produced by photosynthesis may be converted into other substances that the plant needs.

Name TWO such substances.

1. cellulose
2. nucleic acid (2)

A **controlled** experiment, any differences are not due to **mass** of organisms or **volume** of indicator

(f) Some students investigated gas exchange between living organisms and their environment. They used four pieces of pondweed of equal mass, six pond snails of equal mass and equal volumes of indicator.

The table below shows how the colour of the indicator changes in different concentrations of carbon dioxide.

colour changes due to changes in **carbon dioxide concentration** (so don't comment on changes in oxygen concentration!)

Colour of indicator	Concentration of carbon dioxide
red	as in the atmosphere
purple	very low
yellow	very high

obtained from table above

two instructions!

The students set up four test tubes, sealed them and left them for two hours. They are shown in the first column of the table below. Complete the table to state and explain the concentration of carbon dioxide in each test tube after two hours.

N.B. 1 You must describe in terms of photosynthesis and respiration.

2 Plants respire **all** of the time, but this process is masked by photosynthesis in the light.

Test tube after 2 hours	Concentration of carbon dioxide	Explanation
1 Indicator purple  Pondweed in light	VERY LOW	PHOTOSYNTHESIS by pondweed (light available) has used up carbon dioxide faster than RESPIRATION has produced it
2 Indicator yellow  Pondweed in dark	VERY HIGH	RESPIRATION has produced carbon dioxide no light so no removal of carbon dioxide by PHOTOSYNTHESIS
3 Indicator red  Pondweed and snails in light	AS IN THE ATMOSPHERE	A balance exists between PHOTOSYNTHESIS (removes CO_2) and RESPIRATION (produces CO_2) – snails are producing 'extra' CO_2 Compared with (1)
4 Indicator very yellow  Pondweed and snails in dark	EXTREMELY HIGH	Both snails and pondweed RESPIRE (produce CO_2) but no light means no CO_2 removed by PHOTOSYNTHESIS

(6)

(Edexcel) (Total 17 marks)

Self-assessment questions

1 Alice and Belinda are identical twins. Charles and David are non-identical twins. The parents of these four children are Fred and Ethel.

a Which two people have the same set of genes in their cells? (1)

b Which pair of people came from one egg fertilized by one sperm? (1)

c Alice is separated from her sister at birth and brought up in a different country. Does this have an effect on her genes? (1)

d Give two physical characteristics that Alice and Belinda share as the result of genetic inheritance. (2)

e At the age of 20, Alice is 10 kilograms heavier than Belinda. Is this difference due to environmental or genetic factors? (1)

2 Major organs in the human body include heart, lungs, liver, and stomach.

a Draw and label each of these organs on the body outline given below. (5)

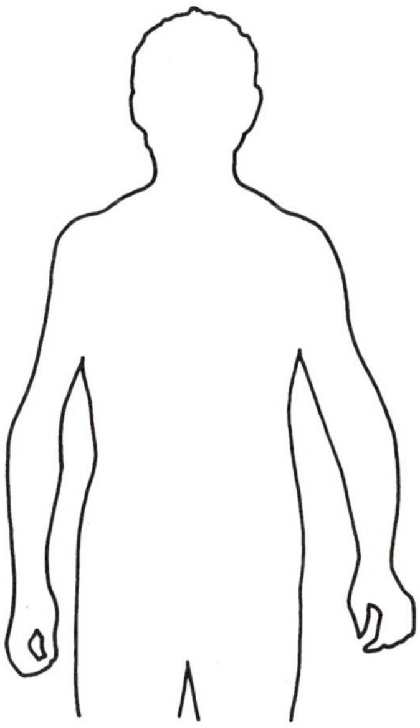

b Describe the function of each of the organs. (4)

c Name an organ that contains valves. (1)

3 Plants use photosynthesis to make glucose from carbon dioxide and water. The rate of photosynthesis depends on light intensity.

a What are the two main uses of glucose in a plant? (2)

b Underline the element (mineral) needed for photosynthesis. (1)

nitrogen phosphorus magnesium potassium sulphur

c Sketch a graph to show how the rate of photosynthesis depends on light intesity. (1)

d Give a factor (other than amount of light, water, or carbon dioxide) that affects the rate of photosynthesis. (1)

4 Your kidneys help to keep constant the amount of water in your blood. Each kidney contains millions of nephrons.

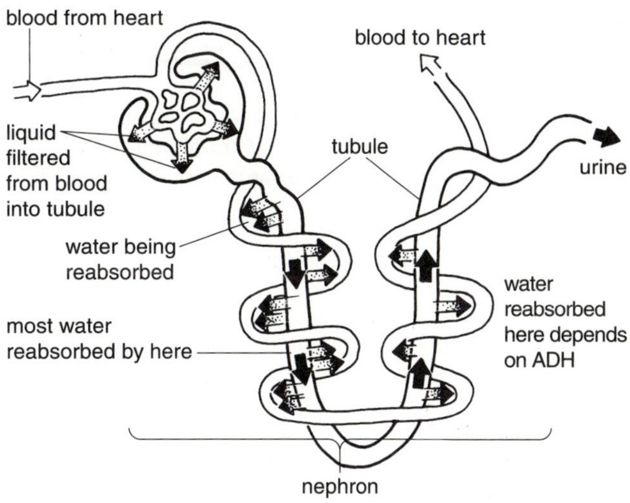

a kidney nephron

a Name another organ that helps to keep constant the composition of your blood. (1)

b Why do you produce less urine in very hot weather? (1)

c The hormone ADH reduces the amount of water lost in urine. What effect does ADH have in a nephron? (2)

d ADH is made in your hypothalamus. Use the idea of *negative feedback* to explain how your pituitary, kidneys, and ADH regulate the water content of your blood. (3)

5 The diagram gives an outline of respiration, the process that releases energy from our food.

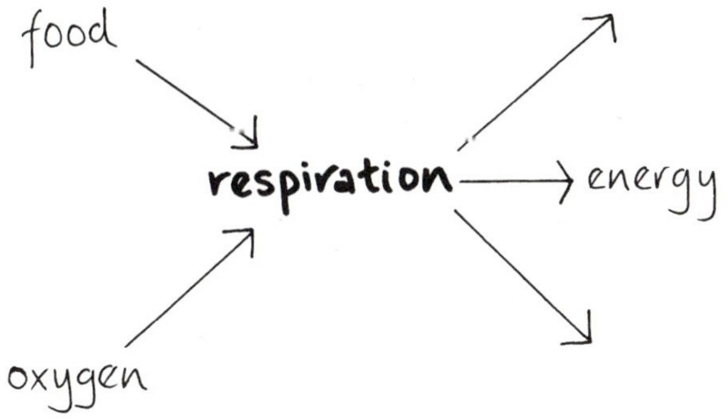

a The results of respiration include two substances. Add them to the diagram. (2)

b Where does respiration take place? (1)

c Name another process that is very similar to respiration. (1)

6 Choose from these words to fill in the gaps in the following sentences:
phenotype genotype allele 1 2 3 4. (4)
Fred has blonde hair, a __________ produced because he is homozygous (hh) for that characteristic. Ethel has dark and is homozygous (HH) for that characteristic. Their son Charles has dark hair. His __________ for hair colour is Hh and the __________ for dark hair is dominant. Charles marries Daisy who has the same genotype for hair colour. There is a 1 in __________ chance that their baby will have blonde hair.

7 Some popular types of apple come from trees that are clones.

a Give one advantage of growing apples from cloned trees. (1)

b Why are the trees not grown from separate seeds? (1)

c Give one disadvantage of growing apples from cloned trees. (1)

8 Here are pictures of five imaginary invertebrates. Preludes and vespers are herbivores that are eaten by both types of fugue. These four invertebrates are all eaten by toccatas.

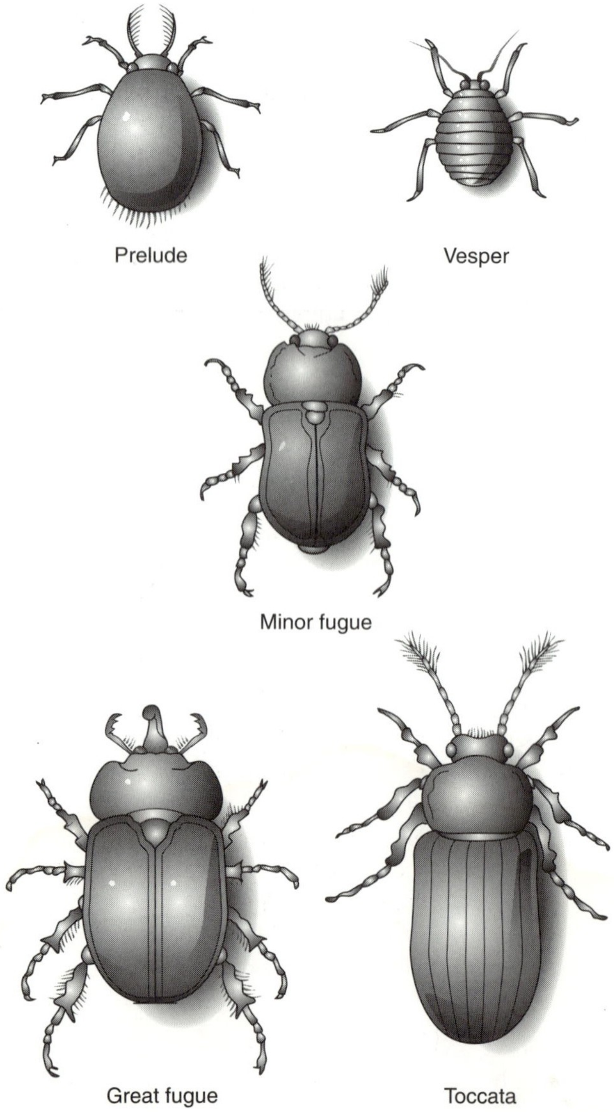

a Make up a key that sorts out these five invertebrates and identifies each one. (5)

b Construct a food chain that shows the feeding relationships of great fugues, plants, preludes, and toccatas. (4)

c Name the producer and the secondary consumer in the food chain in b above. (2)

9 This diagram shows how nitrogen moves between the atmosphere, substances in the ground, and living organisms.

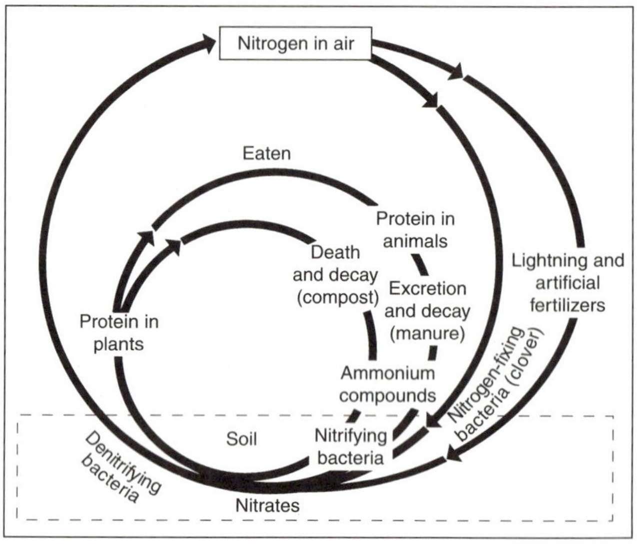

a For a natural woodland habitat, explain why the concentration of nitrate in the soil remains constant. (2)

10 The diagram shows a pyramid of numbers for a community of cabbages, slugs, toads, and a crow that are linked in a food chain.

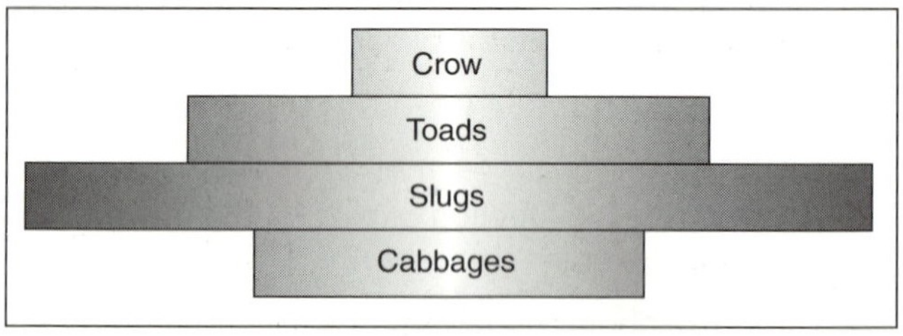

a Draw and label a pyramid of biomass for this community. (2)

b Explain why the pyramid of biomass is more useful than the pyramid of numbers for studying the flow of energy through this food chain. (2)

c Explain why the biomass of slugs is greater than the biomass of toads. (2)

11 The diagram shows the steps involved in mitosis in the nucleus of a cell.

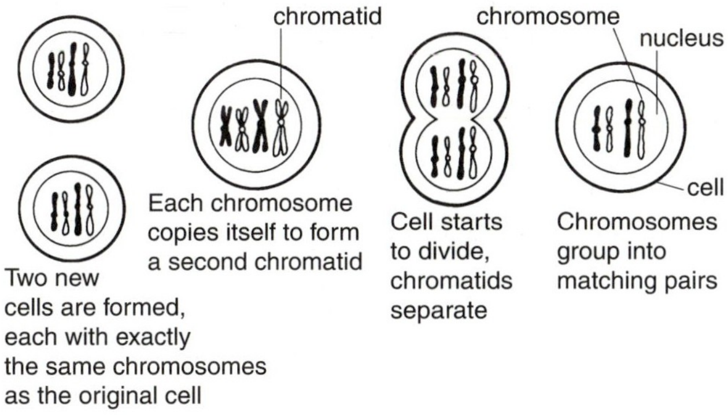

a Number the steps 1–4 to show the correct sequence. (2)

b Egg and sperm cells (gametes) are formed by meiosis. Compare this with mitosis by completing the following chart. (4)

	meiosis	mitosis
number of cells formed from 1 original cell		
number of chromosomes formed from 1 original chromosome		

12 The graph shows how the numbers of plants, slugs, and toads in a habitat vary with time.

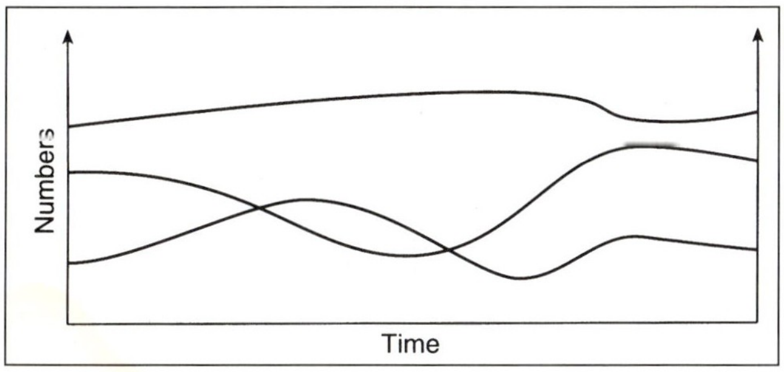

a Label the three curves to show which organism each refers to. (3)

b Give two ways that nutrients return to the soil from toads. (2)

13 Most illnesses are caused either by microbes entering our bodies or by eating an unsuitable diet for many years.

a Which human organ can microbes most easily enter? (1)

b Which human organ is most likely to be damaged by a diet containing too much fat? (1)

c Which two organs are most affected by over-consumption of alcohol? (2)

14 This diagram shows the structure of a leaf.

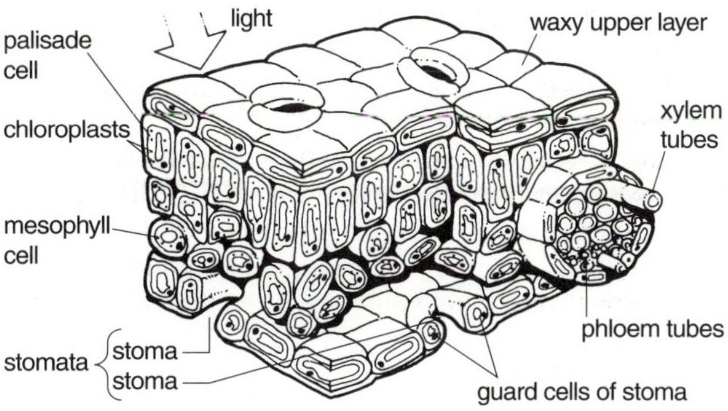

a Add labelled arrows to show where the following substances enter or leave the leaf: carbon dioxide; glucose solution; oxygen + water vapour. (3)

b Why are the cells in the surface of the leaf thin and flat? (2)

15 Seals, seaweed, periwinkle snails, and crabs are members of a food chain.

a Which organism is producer? (1)

b Which organism is the first consumer? (1)

c Arrange the four organisms to show their order in the food chain. (2)

d Why are there more periwinkle snails than crabs? (2)

16 Genetically engineered bacteria can produce hormones needed by humans. Explain the steps involved in altering a bacterium by genetic engineering. (2)

17 Anna is running a 1500 m race. The graph shows how the percentage of anaerobic respiration in her muscles changes during the race.

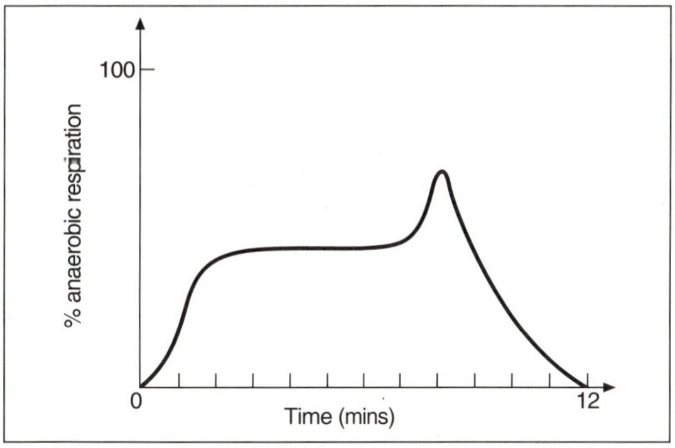

a How long does the race last? minutes (2)

b Sketch the graph expected when Anna runs a 400 m race. (2)

c Explain why Anna breathes heavily for 3 minutes after the race. Include the words: lactic acid; oxygen debt; aerobic respiration. (3)

18 Write a suitable word in each of the gaps in the following passage. Choose from these words: *pumps muscles valves oxygenated deoxygenated lung body chest.* (6)

The human heart is made up from two __________ joined together. __________ make sure the blood flows in one direction only. __________ blood from the body __________ enters the heart which pumps it to the __________. __________ blood returns to the heart which then pumps it around the __________.

19 Explain why the mice in cold countries have thicker coats than the mice in warm countries. In your account, include the words: *natural selection, mutation, variation.* (4)

20 The chart contains organs found in either plants or animals.

ORGAN	PLANT OR ANIMAL?
xylem tube	
pituitary	
pancreas	
stoma	
thyroid	
choroplast	
placenta	
stamen	

a Fill in the chart to show where each organ is found - in a plant or in an animal. (3)

b Underline the organ that you have in your head. (1)

c Ring the organ found in the stem of a plant. (1)

21 The plants and trees in a forest compete with each other for nutrients and light.

a Why do bluebells flower in the Spring before the trees grow their leaves? (1)

b Why do trees grow tall and spindly when crowded close together? (1)

c Give one characteristic that will help compete with others for nutrients. (2)

Answers

(Short answers only)

1 a Alice and Belinda

b Alice and Belinda

c No

d eye colour, tongue rolling

e environment

2 b heart: pumps blood around the body

lungs: gas exchange for respiration

liver: processing absorbed foods and storage and release of useful substances

stomach: stores and mixes foods

c heart

3 a respiration, growth

b magnesium

d temperature

4 a liver

b more water lost through perspiration

c ADH causes more water to be reabsorbed

d see page 27

5 a carbon dioxide, water

b in cells

c photosynthesis

6 phenotype

genotype

allele

4

7 a all apples will be the same

b Because seeds are products of random mixing of parental genes, and so the trees grown from them are not identical to either parent.

c there is no chance of improving the apples

8 b

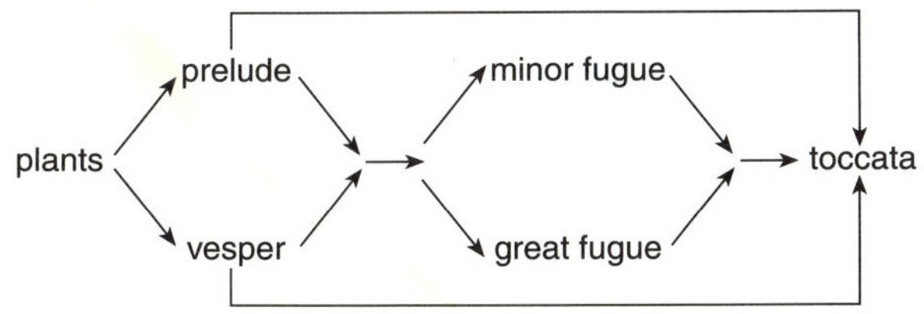

c plants, toccata

9 a Nitrate is removed by plants and denitrifying bacteria at the same rate as it is generated by nitrogen-fixing and nitrifying bacteria.

10 b The pyramid of numbers tells nothing about efficiency of flow.

c Energy transfer between trophic levels is not 100% efficient (it is between 10% and 20%).

11 a 4 2 3 1

b

4	2
2	2

12 a at start:
top: plants
middle: slugs
bottom: toads

b faeces
decay of dead toads

13 a mouth

b heart

c liver, brain

14 b so that light can reach chloroplasts concentrated in central cells

15 a seaweed

b periwinkle snails

c seaweed → periwinkle snails → crabs → seals

d they are smaller and energy transfer to crabs is not very efficient

17 a 10 minutes

18 pumps, valves, deoxygenated, muscles, lungs, oxygenated, body

20 plant
animal
animal
plant
animal
plant
animal
plant

21 a They can get enough energy from the Sun to complete their reproductive phase.

b growth is directed upwards to reach light

c a large root system

INDEX

A

absorption 13
accommodation 23
Acetobacter 53
acid rain 58
acne 31
active site 10
active transport 5, 9, 60, 64
adaptation 55
additives 17
adenine 43, 44
ADH (antidiuretic hormone) 22, 27
ADP (adenosine diphosphate) 5
adrenaline 22
Agrobacterium tumefasciens 49
AI (artificial insemination) 67
albinism 43
alcohol 20, 24, 51, 54
algae 40, 59, 61
allele 42, 44, 45
amino acid 11, 13, 26, 43, 64
ammonium ion 64
amnion 33
amniotic fluid 33
amylase 10, 13, 60
anaemia 11, 12
angina 14
Angiosperm 7, 40
Anopheles 16
anorexia 12
antagonistic pair 30
anther 38
antibiotic 16, 18, 20, 52, 65, 67
antibody 18, 19, 33
antibody, monoclonal 19
antigen 19
antiseptic 16, 60
aorta 14
aphid 35, 63, 66
artery, coronary 14
artery, hepatic 14
artery, pulmonary 14
artery, renal 14, 27
artery, umbilical 33
arthritis 30
arthropod 40
artificial selection 52
athlete's foot 18
atmosphere 58
ATP (adenosine triphosphate) 5, 30
atrium 14
autotroph 55
axes 70
axon 21

B

bacteriocide 18, 52
bacteriostatic 52
bacterium 18, 40, 48, 50, 51, 52, 53, 59, 60, 61, 64
baking 54
barbiturate 20
base pair 43, 44
bee 38
bicarbonate indicator 5
biceps 30
bile 13
biogas 51
biological control 63, 66
bioreactor 51, 53
bladder 27, 31
blind spot 23
blood cell, red 15, 30
blood cell, white 15, 30
blood 15
BOD (biological oxygen demand) 59, 61, 65
body temperature 5, 15, 29
brain 24, 32
bread 54
bronchitis 20
budding 54

C

caffeine 20
calcium 11, 15, 18
cambium 35
cancer, bladder 20
cancer, colon 12
cancer, lung 16
cancer, skin 16, 58
cap (diaphragm) 31
capillary 13, 14, 18
carbohydrate 11, 64
carbon cycle 60, 64
carbon dioxide 5, 15, 26, 30, 33, 34, 51, 54, 58, 64, 66
carnivore 55, 62
carrier proteins 9
carrier 45
carrying capacity 56, 57
cartilage 30
catalyst 10
cell division 5, 42, 43, 44
cell membrane 8, 9, 11, 13
cell, plant and animal 8
cellulose cell wall 8
cellulose 34
centromere 44
cerebellum 24
cerebral cortex 24
cervix 31, 33
CFC (chlorofluorocarbon) 58
cheese 53
chlorophyll 8, 34, 37
chloroplast 8
cholera 16, 17
cholesterol 11, 12
choroid 23
chromatid 42
chromosome 8, 42, 43, 44, 46, 48
ciliary muscle 23
cirrhosis 16, 20
clinistix 10, 25
clone 50
cobalt chloride paper 36
coil 31
colour blindness 46
combustion 64
community 55
condom 31
cone cell 23
constipation 12
consumer 62
contraception 31, 32
control 68
cornea 23
cotyledon 39
crossing over 47
cuticle 7, 36
cystic fibrosis 16, 45
cytoplasm 8
cytosine 43, 44

D

DDT 59
deamination 26
decomposition 62
denaturation 10, 29
denitrification 64
depression 16
diabetes 16, 25
dialysis 28
diaphragm 6
diarrhoea 17, 20
diet, adequate 11
diet, balanced 11
diffusion 9
digestion 13
diploid 42, 46
disease 16
dispersal - wind, water, animal 39
DNA replication 42, 43
DNA 43, 44, 48
dominant 45
double helix 44
Down's syndrome 44
drugs 20, 24
drugs, effect on synapse 21
dysentery 17

E

ecology 55
ecosystem 55, 65, 66, 67
egg 50
embryo 33, 39, 50
emphysema 20
endocrine gland 22
endonuclease 48
endosperm 7
endothermy 6, 13, 29
energy 5, 9, 11, 54, 62
environmental resistance 47, 56, 65, 66, 67
enzyme 5, 8, 10, 48, 53, 60, 64
enzymes - uses 10
epidermis 35
epididymis 31
eutrophication 59
excretion 26, 64
eye 23

F

factor-8 48, 49
fair testing 68
fermentation 54
fermenter 48, 51, 52, 53
fern 40
fertilisation 39, 42, 44, 46, 47, 50
fertiliser 59

fever 29
fibre 11
fibrin 18
fibrinogen 13, 15, 18
filament 38
fish farming 65
fish 40, 65
flaccidity 36
flower 7, 38
foetus 33
food chain 60, 62
food preservation 53
fossil fuel 58, 64
fruit 39
fungus 18, 40, 50, 60, 64, 66

G

gall bladder 13
gamete 31, 39, 42, 44, 46, 47
gasohol 51
gene probe 45
gene transfer 49
gene vector 48
gene 8, 42, 43, 44, 45, 48
genetic engineering 10, 25, 48, 66
genotype 44, 45
genus 41
geotropism 37
global warming 58
glucagon 13, 22, 25
glucose 5, 11, 13, 15, 25, 30, 34, 54
glycogen 8, 13, 25
gonorrhoea 18
Graafian follicle 32
graphs 70
greenhouse effect 58
grey and white matter 21
growth 5, 11, 33, 67
growth, plant 7
guanine 43, 44
gut 13

H

habitat 55
haemoglobin 15
haemophilia 46
haploid 42
hayfever 16
heart disease 12, 14
heart 14
heat 5, 30, 60
hepatitis 16, 19
herbicide 66
herbivore 55, 62, 63
heroin 20, 24
heterotroph 55
heterozygous 45, 46
homeostasis 6, 15, 25, 29
homologous pair 42, 44, 46, 47
homozygous 45, 46
hormone 8, 11, 15, 20, 22, 31, 32, 37
horticulture 66
human growth hormone 22, 48
humerus 30
humidity 36, 66
Huntington's disease 45
hybridoma 19
hydrochloric acid 18
hyperglycaemia 25
hypoglycaemia 25
hypothalamus 24, 27
hypothermia 20, 29
hypothesis 68

I

immunity - active, passive, natural, artificial 19
immunity 16, 17
independent assortment 47
influenza 16, 18
insect 40
insecticide 16, 66
insulin 13, 15, 22, 25, 48
iris 23
iron 11
irradiation 17
IUD (intra-uterine device) 31

J

joint 30

K

karyotype 46
key 40
kidney 26, 27, 28
kingdom 41
kwashiorkor 11, 12

L

lacteal 13
lactic acid 30, 54
large intestine 13
lead 58, 59
leaf 7, 34
lens 23
ligament 30
ligase 48
light 34, 66, 67
limewater 5
limiting factor 34
Linnaeus 41
lipase 10, 13, 60
lipid 9, 11
liver 13, 16, 26
loop of Henle 27
louse 65
LSD 20, 24
lung 6, 14, 26
lymphocyte 15, 18, 19
lysozyme 18

M

magnesium 37
malnutrition 11, 12, 20
mammal 40
mammary gland 6
marasmus 11, 12
mean 68
medulla 24
meiosis 42, 44, 47
membrane, selectively-permeable 9
Mendel's law 45
meninges 24
meningitis 24
menstrual cycle 32
menstruation 32
meristem 50
mesophyll 34
metabolism 8, 13, 15
methane 51, 58, 60, 61
microorganism 51
micropyle 39
mineral ions 9, 11, 34, 37
mitochondria 8
mitosis 42
moss 40
mould 52
mouth 13
movement 5
muscle 30
mutagen 43
mutation 43, 44, 47
myelin sheath 21

N

natural selection 47
negative feedback 25, 27, 29, 32, 56
nephron (kidney tubule) 27
nervous system 6, 21
neurone 21
neurotransmitter 21
night blindness 12
nitrate 37, 59, 64
nitrification 64
nitrogen cycle 60, 64
nitrogen fixation 49, 64
nitrogen 37, 64
nitrogen-fixing bacteria 7
node of Ranvier 21
nucleotide 43, 44
nucleus 8, 43

O

obesity 12
oesophagus 13
oestrogen 22, 32
oil 59
optic nerve 23
osmosis 9
osteoporosis 12
ovary 7, 31, 38
oviduct 31
ovulation 32
ovule 39
oxygen debt 54
oxygen 15, 30, 33, 34, 54, 59, 60
ozone 58

P

painkiller 20
pancreas 13, 25
parasite 63
pasteurisation 17, 53
pathogen 15, 16, 18
penicillin 52
Penicillium 52
penis 31
pepsin 13
peristalsis 13
pest 63
pesticide 49, 59, 63
petal 38
pH 10, 15
phagocyte 8, 15, 18
phenotype 43, 44, 45
pheromone 63
phloem 34, 35
phosphate 59
photosynthesis 34, 62, 64
phototropism 37
pill 32
pituitary hormones 31, 32
placenta 33
plasma 15, 26
plasmid 48, 49
Plasmodium 16
platelets 15, 18
polio 19
pollen 39
pollination 38, 66
pollution 58,59
population 55, 56, 57, 63
positive feedback 59

potometer 36
predator 63
prey 63
producer 55, 62
progesterone 22, 31, 32
prostate gland 31
protease 10, 60
protein 9, 11, 13, 18, 43, 44, 64, 65
protoctista 40
Pruteen 53
pulse 14
pupil 23
pyramid of biomass 62
pyramid of numbers 62
pyruvic acid 5, 54

R

radiation 43, 58
radiotherapy 16
radius 30
receptacle 38
recessive 45
rectum 13
reflex action 24
reflex arc 21
refrigeration 17
rennin 53
reptile 40
resistance 49, 52
respiration 13, 26, 62, 64
respiration, aerobic 5
respiration, anaerobic 30, 54
respirometer 5
restriction enzyme 10, 48
retina 23
Rhizobium 7, 49
ribosome 8, 43
rickets 11, 12
rod cell 23
root 7, 35
root nodule 7, 64
roughage 11, 12

S

Salmonella 16, 17
saprotroph 17, 60
schizophrenia 16
SCP (single cell protein) 53
scrotum 31
scurvy 11, 12, 16
seed 39
selective breeding 65, 66
sepal 38
sewage 59, 60, 61
sex linkage 46
sex 46
shivering 29
sickle cell anaemia 16, 43
skeleton 30
skin 6, 29
sludge 61
small intestine 13
smoking 14, 16
solvent abuse 20
species 41
sperm 31, 50
spermicide 31
sphincter 27
spider 40
spinal cord 21
sporulation 52
stamen 38
starch 8, 11, 13, 34
stem 35
sterilisation 17, 31, 50
stigma 38
stomach 13
stomata 7, 34, 36
style 38
sucrose 11, 34
surrogacy 50
sweating 29
synapse 21
synovial fluid 30

T

tables 69
tendon 30
testis 31
testosterone 22, 31
thymine 43, 44
tobacco 20
tooth decay 12
toxin 17
transcription 43
transgenic organism 49
translation 43
transpiration 36
transplant 28
triceps 30
turgidity 8
typhoid 17

U

ulna 30
umbilical cord 33
undernutrition 11
units 69
urea 15, 26, 27, 33, 64
ureter 27, 31
urethra 27
uterus 31, 32, 33

V

vaccine 19
vacuole 8
vacuole, contractile 8
vagina 31
valve 14
valve, cuspid 14
valve, semi-lunar 14
variables 68
variation 47
vas deferens 31
vascular bundle 35
vasectomy 31
vasoconstriction and vasodilation 29
vegetative propagation 50
vein, hepatic portal 14
  hepatic 14
  pulmonary 14
  renal 14, 27
  umbilical 33
vena cava 14
ventricle 14
vesicle 8
villus 13, 33
vinegar 51, 53
virus 18, 33
vitamins 11
vulva 31

W

water potential 9, 15
water 11, 15, 34, 36, 60, 64
weedkiller 37
wilting 36

X

X chromosome 46
xylem 7, 9, 34, 35, 36

Y

Y chromosome 46
yeast 54
yellow spot (fovea) 23
yoghurt 53

Z

zygote 33, 39, 42, 50